GEOBOTANY

GEOBOTANY

Edited by

Robert C. Romans
Bowling Green State University
Bowling Green, Ohio

PLENUM PRESS · NEW YORK AND LONDON

Library of Congress Cataloging in Publication Data

Geobotany Conference, Bowling Green State University, 1976.
  Geobotany.

  Includes bibliographical references and index.
  1. Paleobotany—North America—Congresses. 2. Botany—North America—Congresses. 3. Phytogeography—Congresses. I. Romans, Robert C. II. Title.
QE935.G46 1976   561   76-51249
ISBN 0-306-31007-4

Proceedings of the Geobotany Conference held at Bowling Green State
University, Bowling Green, Ohio, February 21, 1976

© 1977 Plenum Press, New York
A Division of Plenum Publishing Corporation
227 West 17th Street, New York, N.Y. 10011

Printed in the United States of America

Preface

The papers in this volume were presented at the Geobotany Conference held at Bowling Green State University, Bowling Green, Ohio, on 21 February 1976. Though such diverse topics as anthropology and paleobotany are covered, all papers utilized the concept of geobotany as a unifying theme.

Nearly a decade ago, the first in this series of geobotany conferences was organized on this campus by Dr. Jane Forsyth of the Department of Geology. After considerable growth, culminating in an International Geobotany Conference at the University of Tennessee in 1973, it was decided to again organize a regional geobotany meeting. The melange of papers in this volume are products of that meeting.

Geobotany, by definition, is an interdisciplinarian approach to interpretational problems involving such investigators as geologists and botanists, archaeologists and stratigraphers, ecologists and palynologists. Interaction among these individuals is necessary for the satisfactory solution of a problem. Each can provide invaluable assistance to the other. The purpose of the meeting in Bowling Green was to provide a forum for the exchange of ideas and information.

Sponsors of the conference include the Department of Biological Sciences, the Department of Geology, the Environmental Studies Center, the College of Arts and Sciences, and the Graduate School. All of the sponsors are academic or administrative units of Bowling Green State University and each played an important role in the success of the conference.

Bowling Green   Robert C. Romans
1976

Contents

LATE PLEISTOCENE AND POSTGLACIAL PLANT COMMUNITIES OF THE GREAT LAKES REGION

Ronald O. Kapp

Alma College

Alma, Michigan   48801

ABSTRACT

The few available interglacial pollen studies from the southern Great Lakes region show developmental sequences and pollen assemblages with striking similarities to postglacial records. A five-stage cycle of glacial and interglacial vegetation phases is postulated and applied to data from this region. There is limited evidence, however, of late Pleistocene changes in genetic potential of certain species (e.g., Picea mariana). Also, some interglacial and interstadial (glacial age) pollen records indicate that ranges of certain tree genera or forest communities extended beyond their modern ranges.

Late-glacial (post-Wisconsinan) records in the glaciated Great Lakes region do not consistently give evidence of tundra vegetation; instead, "open forests" may have existed in ice-margin areas. The major climatic/ecologic break at about 10,500 B.P. initiated a developmental sequence of postglacial hardwood forests which is remarkably consistent throughout the region. Migration routes were influenced by both physiographic (Appalachians, Great Lakes) and climatic factors. The spread of Fagus and Tsuga to their present ranges in glaciated areas is re-examined; additional detailed radiocarbon-dated analyses are required to document these migrational patterns.

The strong correspondence of patterns of contemporary pollen spectra and vegetational distribution has been revealed by recent detailed studies; this restores confidence in the paleoecological validity of pollen analysis. By utilizing statistical procedures, both regional and local vegetational composition can, within limits,

be directly correlated with pollen spectra. Overall, paleoecological research suggests that plant communities of the Great Lakes
region have had remarkable continuity throughout the late Pleistocene.
Although climatic conditions, extensive glacial meltwater channels,
and fluctuations in the level and drainage of the Great Lakes apparently explain certain extensions of range and novel plan assemblages, patterns of forest migrations and succession are predictable.

INTRODUCTION

It is especially appropriate, and a personal privilege for me,
to include a review of palynological and related evidence of Pleistocene vegetation in the proceedings of an Ohio geobotany conference.
Ohio is the ancestral home of North American geobotany, especially
of pollen analysis. Paul B. Sears published his first paper on
postglacial paleoecology in 1930, only 14 years after the emergence
of the fledgling science in Europe (von Post, 1916). The early work
of Sears in Ohio (1930, 1931) and John Voss in Illinois (1933, 1934)
prompted Sears to publish the first review paper on this subject in
1935, over 40 years ago! This apparently stimulated an interest in
palynological research.

The prodigious research of Sears (1935b, 1940, 1941, 1942a,
1942b, 1942c, 1951a, 1952, 1955, 1956) and Potzger (1945, 1946, 1948,
1951, 1952, 1953) and their associates (Sears and Clisby, 1952; Sears
and Bopp, 1960, Potzger and Courtemanche, 1956a, 1956b; Potzger and
Otto, 1943) led to a series of more comprehensive review papers on
the history of vegetation of the Great Lakes and Eastern North America by Sears (1948, 1951b), Deevey (1949, 1957), Martin (1958a), and
Just (1959), all prior to 1960.

Beginning in about 1960, a period of logrhythmic growth in the
quantity and sophistication of research in both glacial geology and
palynology began. Extensive research programs were initiated by
Wright in Minnesota; Bryson in Wisconsin; Benninghoff and Margaret
Davis in Ann Arbor; Frey, Gooding, and Wayne in Indiana; Goldthwait,
Forsyth, and Ogden in Ohio; as well as Terasmae, Dreimanis, and their
associates in Ontario. These studies of the past 15 years have confirmed the basic patterns revealed by the work of the early researchers. Moreover, the availability of radiocarbon determinations,
studies of contemporary pollen dissemination and its statistical
correlation with vegetation and with fossil pollen spectra have improved our understanding of the nature of late-glacial and postglacial vegetation. Reviews of these recent developments by Ogden
(1965), M.B. Davis (1965), Cushing (1965, 1967), and Wright (1964,
1968, 1971) suggest that we may already have been surfeited by an
abundance of overviews and syntheses. Despite these misgivings, a
review of the evidence for late Pleistocene vegetation seems to be an
appropriate part of this volume on the geobotany of the Great Lakes
region.

The scope of this review is the last half of the Pleistocene. Although evidence is still sketchy, we can now begin reconstruction of vegetation of the second (Yarmouthian) and third (Sangamonian) interglacial stages in the southern Great Lakes region. Vegetation of the related glacial stages can only be determined by studies of deposits at ice-free sites (pre-glacial, interstadial, late-glacial) in the glaciated region or at depositional localities beyond the glacial border. The sequence of Pleistocene stages and substages to which reference is made is given below (youngest to oldest):

(Stages)   (Interstadials-Erie Lobe)

Holocene

                                    Late glacial

              Main (Late)

                                  Plum Point (= Sidney ?)

Wisconsinan glacial Middle

                                  Port Talbot I & II

                                  St. Pierre

              Early

Sangamonian interglacial stage
Illinoian glacial stage
Yarmouthian interglacial stage
Kansan glacial stage

PRE-WISCONSINAN GEOBOTANY

Yarmouthian Interglacial Record

The Illinoian glacial drift is sufficiently thin in certain areas of southeastern Indiana that underlying Yarmouthian and Kansan deposits are exposed in stream cuts (Gooding, 1966). At the Handley Farm section in Fayette Co., Indiana, a stream bank exposure includes 8 m of pollen-rich lacustrine clay; these interglacial deposits have Kansan till beneath and are overlain by pro-Illinoian glacial outwash and Illinoian till. The pollen record (Kapp and Gooding, 1974) is dominated by deciduous species; the pollen diagram (Fig. 1) is divided into three pollen zones:

1. The Oak-Ironwood Pollen Zone is characterized by high Ostrya-Carpinus pollen (20-35%), Quercus (about 30%), and lesser amounts of Picea, Pinus, Tilia, Corylus and Populus.
2. The Oak-Elm-Beech Zone is characterized by decrease in Ostrya-Carpinus pollen and increases in Ulmus (to 20%), Fagus (to 20%) Celtis, Carya, Fraxinus and Acer; conifers, at least Pinus and Picea, were not in the region.

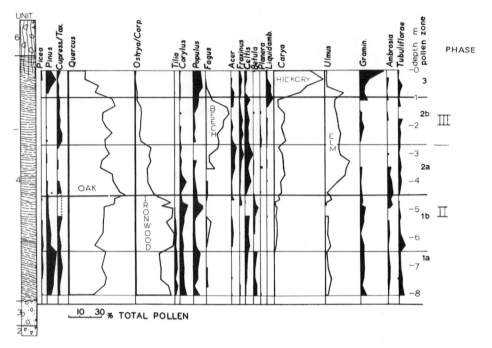

Figure 1. Yarmouthian interglacial pollen diagram from Handley Farm section, southeastern Indiana (adapted from Kapp and Gooding, 1974, p. 231).

3.   The Sweetgum-Hickory Zone is characterized by a peak in Carya (50%), high frequencies of Liquidambar, Populus and Gramineae pollen, increased conifer representation and sharp declines in frequencies of Quercus, Fagus and Ulmus.

The sequence of development of this deciduous tree pollen record is strikingly similar to that seen in many postglacial pollen diagrams; commonly the deciduous types are represented by successive maxima in the sequence: Ostrya/Carpinus (usually with Betula)⟶ Ulmus ⟶ Fagus/Acer ⟶ Carya (or Quercus, Gramineae, etc.) It is likely that this sequence is the result of several interlinked factors including: speed of migration, ease of establishment on youthful sites, climatic requirements and tolerance, ecological and successional characteristics. The hypothesis that the sequence of appearance of major pollen types is virtually the same throughout the deciduous forested sections of the Great Lakes region, both in interglacial and postglacial records, requires further testing. The Handley Farm diagram suggests that such may be the case.

In a yet broader generalization, it was postulated by Turner

and West (1968) that a four-phased pattern typifies European interglacial sequences; Wright (1972) applied this model to eastern North
America. The four phases are:

      I.   Late-glacial or pre-temperate phase
     II.   Early temperate phase (maximum warmth)
   III.   Late temperate phase
    IV.   Post-temperate (or pre-glacial) phase

For unglaciated sites, V. Full glacial phase, can be added.

The Yarmouthian pollen diagram from the Handley Farm site seems
to include interglacial Phases II and III; persistence of low frequencies of pine and spruce pollen into Early temperate phase gives
residual evidence of the end of the preceding Late-glacial phase.
The reappearance of these two conifer genera near the top of Phase
III reflects the beginning of climatic deterioriation of the Late-
temperate phase.

Illinoian Glacial/Sangamonian Interglacial

At Pittsburg Basin in south-central Illinois, about 60 km south
of the limit of Wisconsinan glaciation, E. Grüger (1970, 1972a,
1972b) analyzed a pollen sequence which includes all five vegetational phases (Fig. 2). The deposits, formed in a lake basin, are
interpreted to extend from late-Illinoian to the present, with the
uppermost horizon disturbed by plowing.

The record begins with Phase I, the Illinoian late-glacial
phase, dominated by Picea (20-50%) and Pinus (15-35%) and including
low frequencies of pollen of hardwood genera or of herbs. This is
succeeded by Sangamonian interglacial phase II which is initially
dominated by pollen of Quercus, Ulmus, Carya, Fraxinus and the
Taxodiaceae/Cupressaceae groups and subsequently reflects the glacial
maximum in the invasion of prairie species (Ambrosia and Gramineae).
Phase III shows a reversion to dominance by hardwood forests. In-
terestingly, Liquidambar is not prominent in the record until this
later phase, supporting the hypothesis that sweetgum requires mature,
well-leached soils. Phase IV (Post-temperate) is probably only fragmentary due to an apparent hiatus in the record. The full-Wiscon-
sinan record (Phase V) at the Pittsburg site seems to give a detailed
record of the demise of the interglacial plant communities and their
replacement by full-glacial vegetation. The mesophytic deciduous
species (Juglans, Liquidambar, Platanus, Fagus, Ulmus, and even Carya)
declined rapidly in south central Illinois at the end of the Sangamon.
Deciduous forest communities were apparently supplanted again by
prairie (high frequency of Artemisia, Ambrosia, other Compositae,
Chenopodineae, Gramineae, and Cyperaceae) with scattered oak-hickory
stands during the Altonian (early Phase V) substage 75,000 - 28,000

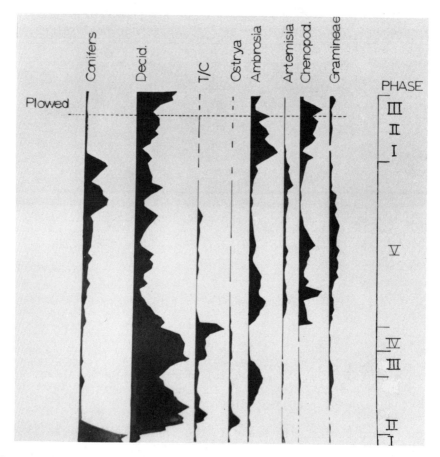

Figure 2. Pollen diagram from the Pittsburg Basin, Fayette Co., Illinois. Records begin in the late Illinoian glacial age (Phase I) and extend through an entire interglacial-glacial sequence into the postglacial (adapted from E. Grüger, 1972a).

years ago. During the Farmdalian (28,000 - 22,000 B.P.) and Woodfordian (22,000 - 12,500 B.P.) substages pine-hardwood and finally spruce-pine-hardwood communities developed.

The uppermost (Phases I, II, III) zones at the Pittsburg basin reflect the expected postglacial sequence for that area, dominated by prairie species with oak/hickory stands. It is significant that the postglacial of south-central Illinois was less favorable for deciduous forests than the equivalent period of the early Sangamonian interglacial. The climate of Phase II of the interglacial seems to have been sufficiently equable to permit establishment of the species-

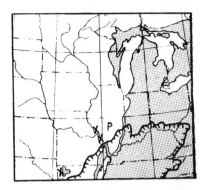

Figure 3. Location of
Pittsburg Sangamonian interglacial site (P) in relation
to the present ranges of
Fagus (gray) and Liquidambar
(⌒⌒×).

rich Western Mesophytic forest
communities (sensu Braun, 1950)
in that region; this is reflected
by the extension of the ranges
of Liquidambar and Fagus into
south-central Illinois (Fig. 3).

Sangamonian pollen records
have also been described from
southeastern Indiana (Englehardt,
1962, Kapp and Gooding, 1964b)
and from the Don Beds at Toronto
(Terasmae, 1960). The Darrah
and Smith Farm sections near
Richmond, Indiana record the
end of Late-glacial Phase I
and Early-temperate phase II.
The record is then truncated
and overlain by involuted sediments which may represent scouring
and disturbance by Wisconsinan

ice. Dreimanis (1973) suggests that these uppermost conifer-pollen-
rich sediments represent the Early Wisconsinan St. Pierre interval;
they clearly correspond to vegetation of Full glacial Phase V.

While the validity of suggesting direct continuity between segments of pollen diagrams from such distant locations as southeastern
Indiana and Toronto, Canada is most questionable, a diagrammatic
representation of these two Sangamonian interglacial records is
suggestive of the probable sequence. Figure 4 is such a composite
diagram. The Smith Farm interglacial sequence from the Whitewater
Basin of Indiana is shown at the base of Fig. 4 and, higher in the
diagram, the glacial segments of this record is plotted.

The pollen record from the Don Beds of Toronto seems to record
interglacial phases III and IV, beginning during the interglacial
climatic optimum and subsequently reflecting the climatic deterioration and floristic changes which occurred with the approach of the
Wisconsinan glacial stage. Terasmae (1960) concludes that the Don
Beds assemblage "represents a mean annual temperature of 5° (3° C)
warmer than the present."

Certain phytogeographical and autecological inferences can be
drawn from these interglacial records. The presence of Liquidambar
pollen in abundance in the Don Beds Sangamonian record is a substantial extension of range (Fig 5). Its presence along with a more
southerly assemblage of deciduous species (Chamaecyparis, Maclura,
Robinia, and Quercus muhlenbergii and Q. stellata) indicates a substantially different forest distribution at the Sangamonian maximum
compared with present. I have already commented on the extension of

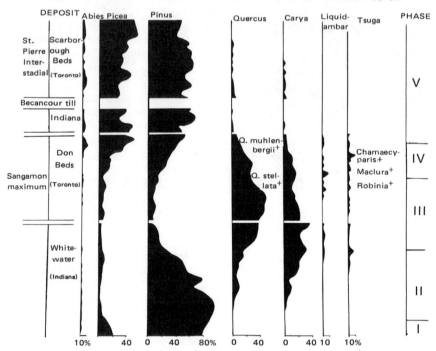

INTERGLACIAL & INTERSTADIAL VEGETATION – S.E. GREAT LAKES REGION

Figure 4.   Composite Sangamonian pollen diagram from Smith Farm section (Wayne Co., Indiana) and Don Beds of Toronto. Macrofossil remains from the Don Beds, originally reported to Coleman (1933) include several species (marked +) whose ranges do not now extend as
far north as Toronto.

the Western mesophytic assemblages (including sweetgum) northwest
into south-central Illinois.

   The Sangamonian record from Smith Farm in southeastern Indiana
includes another autecologically significant feature. Picea (believed on the basis of size to be P. mariana) pollen persists in
low but significant percentages throughout the interglacial record
(see Fig. 6 for present range). I postulate that this reflects
broader ecologic amplitude in that species during the last interglacial than is true of postglacial populations in the Great Lakes region. Postglacial pollen diagrams from the southern Great Lakes
region have virtually no Picea pollen after 10,000 years B.P. The
genotypic diversity and ecotypic flexibility of that species seems
to have been reduced during the Wisconsinan. We should look for
evidence of this type of biotype depauperation as pollen-analytical
studies are pushed back into the earlier stages of the Pleistocene.
Unfortunately, only very few pollen-rich interglacial deposits have

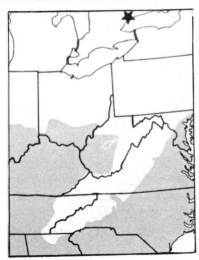

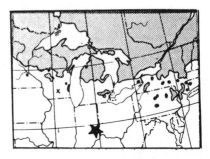

Figure 5.   Modern distribution
(gray) and Sangamonian interglacial record (star) of Liquidambar.

Figure 6.   Modern distribution
(gray) and Sangamonian interglacial record (star) of Picea
mariana.

been discovered in North America and only the few described above are
in the Great Lakes region. It is imperative that the vegetational
patterns postulated above be tested at numerous additional sites.

WISCONSINAN FULL-GLACIAL AND INTERSTADIAL VEGETATION

The ecological conditions and vegetational distribution of the
full-sized glacial stage is perhaps the most poorly understood, yet
most widely described and misrepresented aspect of Pleistocene biogeography. Synthesizers from Darwin (1859) to Dorf (1959) and Martin
(1958) have tended to present an oversimplified, but logically ap-
pealing portrait of full glacial vegetation in which broad belts of
climate/vegetation were displaced southward by the advancing continental glaciers of eastern North America. These simplistic theories
have too often found their way into ecology, geology and biogeo-
graphic textbooks.

Scattered evidence of white spruce and other northern species
from such southern latitudes as Texas and Louisiana have served at
once to both support these hypotheses and cause consternation regarding the means of survival of the floristically-rich eastern deciduous
forests. The question posed by Deevey (1949) nearly 30 years ago
still stands: where were the refugia for deciduous species, short
of extensive migrations to southern Florida or central America? Many
paleoecologists have come to explain the situation as a dynamic mix-
ing of northern and southern species under stress of Wisconsinan continental glaciation.

It is beyond the scope of this paper to review all evidence of full-glacial vegetation for eastern North America; Whitehead (1973) and Watts (1970) have provided the most recent treatises on that subject. Brief examination of evidence for southern distribution of northern species, such as are now found in the Great Lakes region, will, however, give some clues regarding the fate of these northern communities during the full-glacial period. In the southern Appalachian Mountains, (northwestern Georgia, Watts, 1975) and at Singletary Lake, North Carolina (Whitehead, 1967) beech (<u>Fagus grandfolia</u>), along with a diverse mixture of mesophytic forest species may have coexisted with pine and spruce in mixed stands or at nearby valley and crest sites. It is unlikely that beech, or the mixed mesophytic forests, occupied sites in the southern Piedmont, Coastal Plain or Florida.

The well-known records of white spruce (<u>Picea glauca</u>) cones and tamarack (<u>Larix</u>) wood in the Tunica Hills region of Louisiana (Brown, 1938) and of spruce pollen from deposits in Texas (Potzger and Tharp, 1954; Graham and Heimsch, 1960) have focused attention on the Mississippi Valley as a Pleistocene plant migration route. Recent re-examination of the Muscotah Marsh sediments of northeastern Kansas by J. Grüger (1973) confirms the abundant presence of <u>Picea</u> pollen (originally misidentified as <u>Abies</u> by Horr, 1955) from 24,000 B.P. to as late as 12,000 B.P. This study and those of the Pleistocene spring deposits of western Missouri (Mehringer, <u>et al</u>., 1968, 1970; King, 1973) all verify the southern distribution of spruce during the Wisconsinan maximum (Main Wisconsinan). A considerable southerly extension of range of spruce along the Mississippi and its tributaries, and in the Ozark Mountains, is now undisputed. In a series of provocative papers Hazel and Paul Delcourt (1974, 1975, in press) have extended the paleoecological studies of the Tunica Hills area of Louisiana/Mississippi. They verify the existance of <u>Picea glauca</u> and <u>Larix laricina</u> macrofossils in these terrace deposits in association with oak, ash, hornbeam, elm, hickory, butternut (<u>Juglans cinerea</u>), black walnut (<u>J. nigra</u>), beech, cucumber magnolia (<u>Magnolia acuminata</u>), <u>Liriodendron</u>, sugar maple, birch, alder and pine. They contend that the dissected loess-covered blufflands which form a high ridge east of the Mississippi Valley served as a migrational route and glacial refugium for both northern (Great Lakes) plant assemblages and for the mixed mesophytic forests. The latter plant communities are also postulated to have occupied sites along other north-south trending streams of Mississippi and Alabama.

Figure 7 shows the modern distribution of black spruce (<u>Picea mariana</u>) in the Great Lakes region (white spruce, <u>P. glauca,</u> does not extend as far south), and reflects the Appalachian distribution of red spruce (<u>Picea rubens</u>). Extreme southerly and westerly Pleistocene localities are plotted; the spruce pollen records off the Florida coast (Davis, 1946), in Texas, Missouri, and Kansas are likely to include both <u>Picea mariana</u> and <u>P. glauca,</u> although presence of the former is paleoecologically more conservative since black spruce now

Figure 7. Modern and late Pleistocene distribution of spruce in eastern North America; modern P. mariana distribution is shown by hachured line and x's (disjunct).

exists in relict stands in cold, acid bogs south of its continuous range. White spruce was clearly present as far south as the Tunica Hills blufflands (although the cones are exceptionally large and the wood somewhat atypical). A recent excavation for a construction project on Nonconnah Creek in metropolitan Memphis, Tennessee, at the site of a mastodon recovery, yielded sediments with a pollen record dominated by spruce pollen with an admixture of deciduous types. At this site, abundant white spruce cones, seeds, branchlets, needles and wood have also been recovered (personal communication H. and P. Delcourt).

In the Great Lakes region itself there is scattered evidence of the character of glacial-age plant communities, primarily in interstadial deposits sandwiched within layers of glacial till or outwash.

Whether reference is to the plant remains of the extensive mid-Wisconsinan Port Talbot interstadials of southern Ontario (Dreimanis, et al., 1966) or to the later Plum Point (= Sidney ?) interstadial of Erie Lobe region, the vegetation is in virtually every instance indicative of a glacial climate. Most interstadial deposits are dominated by conifer pollen and herbaceous (NAP), usually predominantly sedge, pollen.

The interstadials of the eastern Great Lakes region have abundant pine (to 70%) and spruce with varying admixtures of thermophilous deciduous species (Berti, 1975). Farther west and south pine pollen is less frequent, but pollen of deciduous species is commonly present in significant frequency, especially in late-glacial deposits. Wright (1968,1971) has pointed out that pine pollen is found in low quantities in the late-glacial of the southern Great Lakes and in Minnesota. It should be remembered that the mid-Wisconsinan interstadials (e.g., Plum Point) were ice-free periods of as long duration as the entire Holocene. The climate during these interstadials seems to have remained essentially glacial (boreal), although relatively high deciduous and pine pollen frequencies in

certain Ontario deposits lead Berti (op. cit.) to envision vegetation ranging from mixed coniferous-deciduous, or even oak-dominated
forests, to boreal and forest-tundra border situations. These mid-
Wisconsinan interstadials were long enough to permit migration of
pine into central Michigan (Miller, 1973). By way of contrast
late-glacial deposits in Michigan and beyond have low pine values.

The Tundra Question

The character of ice-margin vegetation in the Great Lakes region has long been debated by paleoecologists, yet definitive data
to establish presence of a tundra zone is still elusive. It is clear
that there were tundra elements in Pennsylvania; but it was more
likely a forest-tundra environment than a regionally-extensive
tree-less landscape (Martin, 1958; Berti, 1975). In the central
Great Lakes, careful search of lowermost sediments for macrofossils
or pollen representing tundra elements has been disappointing. West
(1961) postulated that the climate of the midwest allowed for growth
of trees near the ice front; midwestern late-glacial pollen spectra
characteristically have high spruce percentage with NAP represented
by substantial Compositae (especially Artemisia) pollen. In New
England this zone gives more conclusive evidence of tundra elements,
and usually has heavy representation of Ericaceae pollen rather than
Compositae. Deevey's (1949) hypothesis that the response of vegetation to the North American glacial front in low latitudes (40° N)
may have permitted forest survival in an ameliorated climate, especially when compared with the more extreme conditions of higher
latitudes in Europe, appears to be valid. The most noteworthy Great
Lakes area record of tundra vegetation is at the Cheboygan Bryophyte
bed (Miller and Benninghoff, 1969). This interstadial deposit,
dating from sometime in the 13,300 - 12,500 B.P. Cary-Port Huron
interstadial, yielded pollen spectra dominated by NAP and macrofossils of such tundra elements as Carex supina, Dryas integrifolia,
Salix herbacea and Vaccinium uliginosum var. alpinum. This is evidence that communities like those at present in the low artic or
tundra-boreal forest ecotone were formed, at least locally, in the
region.

The Wisconsinan glacial stage ended with remarkable abruptness
and regional synchroneity about 10,500 years ago (Ogden, 1967). This
was apparently due to a significant climatic warming; glacial ice
had retreated by this date from the Great Lake basins. Pollen diagrams throughout the region record a sharp decline in the frequency
(both relative and absolute) of spruce pollen at that date; this
ushered in the plant migrations and vegetational development of the
Holocene (postglacial) plant communities.

HOLOCENE VEGETATION

Pollen analysts have by now published dozens of postglacial
pollen diagrams, some very fragmentary, others in great detail and
extending through the Holocene period. Following the spruce decline
at about 10,500 B.P., a rather predictable sequence of pollen maxima
can now be anticipated beginning with a pine period, followed
by peaks for Ostrya-type (usually with Fraxinus and Betula), Ulmus,
Quercus, and finally Fagus-Acer or possibly Carya. Figure 8, a
simplified pollen diagram (unpublished) from Demont Lake in central
lower Michigan, will serve to illustrate the general late-glacial
and postglacial sequence typical of the region.

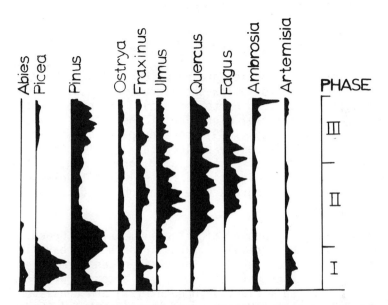

Figure 8.   Pollen diagram from Demont Lake, Isabella Co., Michigan
(Ahearn, unpublished), showing a typical late-glacial and post-
glacial sequence of pollen maxima. The Pre-temperate (I) Early-
Temperate (II) and Late-Temperate interglacial phases are represented.

The late glacial spruce-herb zone is properly considered the
interglacial Phase I (Pre-temperate phase). Pine pollen (likely
of distant origin), and certain thermophilous trees are in most every
such diagram found in the spruce zone. We believe that an open
spruce forest (with white spruce dominant) subsequently became an
extensive closed stand (both white and black spruce). Fraxinus

nigra was apparently the species of ash to dominate the late-glacial
zones, with white ash (four-furrowed pollen) predominant later.
While Pinus pollen is overrepresented in the diagrams, compared
with its frequency in the forest, it increased sharply to form an
ephemeral pine pollen zone at about 9,500 B.P.

The sequence of entry of the species of Tsuga, Fagus, and Tilia
into the pollen diagrams varies slightly from the southeastern to
the northwestern Great Lakes region (Wright, 1964, Cushing 1965).
These variations are probably the result of differing distances of
the study sites from the full-glacial refugia for the species in
question. Rates of migration and migrational routes (controlled by
ecological and physical barriers) certainly influenced the development of the modern forest communities.

Fagus and Tsuga Migration

The distinctive pollen grains of beech (Fagus) and hemlock
(Tsuga) can be used to follow the migration routes of these species
(monospecific in eastern North America). The opportunity to reconstruct the postglacial migration of these two species was long
ago discussed by Deevey (1949) and many palynologists have been
conscious of, and frequently radiocarbon-date the time of, influx
of these pollen types of postglacial diagrams. While it is still
premature to plot the migrational routes definitively, enough radiocarbon-dated pollen diagrams have been published to formulate a
working hypothesis.

Figure 9 is a proposed migrational model for beech. Radiocarbon dates on the arrival of Fagus into an area have in some
instances been extrapolated on the basis of uniform rate of
sedimentation; these must, of course, be accepted with caution.
Fagus pollen was sufficiently abundant (2%) and consistently present
in central Ohio at the beginning of the pine period (Ogden, 1966)
to suggest a nearby refugium and a migrational sequence beginning
in Ohio/West Virginia/Pennsylvania about 10,000 B.P. Beech seems to
have moved rather slowly into central Ohio and northern Indiana for
the estimated dates of its establishment at Sunbeam Prairie Bog
(Fig. 8, No. 19) is 8,500 B.P., 7,600 B.P. at Pretty Lake (No. 13)
and 8,000 B.P. at Fox Prairie Bog (No. 17); Fagus pollen was insig-
nificant at the Wells Mastodon Site (No. 15) up until the time deposition ceased at 3,000 - 5,000 B.P. (Kapp and Gooding, 1964; Ogden,
1969; Engelhardt, 1965, Gooding and Ogden, 1966). It is quite possible that the eastward extension of the "prairie peninsula" 8,000 -
4,000 years ago (Wright, 1968) may have retarded, or even prevented
the invasion of Fagus to the western limit of its range in Indiana
until the climatic cooling of very recent time (Benninghoff, 1963).
In fact, Fagus may even have entered extreme northwestern Indiana
(Hudson Lake, No. 14, date of arrival 5,700 B.P.) from the north.

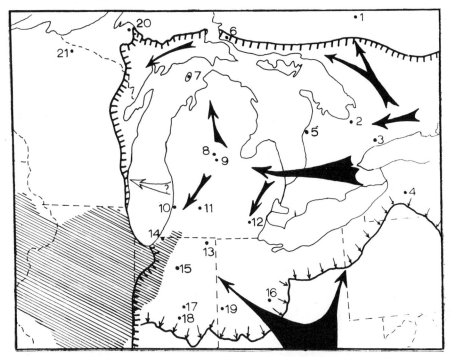

Figure 9. Suggested postglacial migrational pathways of American beech (<u>Fagus</u> <u>grandifolia</u>). The hachured line is the distributional limit of the species; the line with arrows shows the approximate southern boundary of Wisconsinan glaciation; prairie is shaded. (See text for site locations and references.)

      In sharp contrast, <u>Fagus</u> seems to have spread rapidly in a northerly direction, possibly entering Ontario across the Lake Erie basin which was almost entirely drained during the period 12,500 - 4,000 B.P. (Forsyth, 1973). <u>Fagus</u> enters the Battaglia Bog region, NE Ohio (Shane, 1975; not plotted on Fig. 8) before 9,000 B.P. and at Van Nostrand Lake north of Toronto (No. 3, McAndrews, 1973) by 9,500 to 9,000 B.P. The extrapolation of dated horizons at Protection Bog (No. 4; Miller, 1973) leads to an estimated date of <u>Fagus</u> arrival of 8,000 B.P. While Karrow <u>et al</u>. (1975) do not provide definite dates, <u>Fagus</u> seems to have arrived at the Lake Louis area of northwestern Quebec (No. 1), beyond the present distribution of beech by approximately 7,300 B.P. (Vincent, 1973). One may speculate about migrational routes from the east but dates of beech influx in New England pollen diagrams are generally younger than those of Ohio. The possibility of a migrational pathway through the mountains and east of Lake Ontario needs to be investigated. It is postulated that beech entered lower and upper Michigan from the east both south and north of Lake Huron. At Quadrangle Lake near Sault Ste. Marie,

Ontario (No. 6) Terasmae's (1967) diagram suggests a date (extrapolated) of beech entry of 6,700 B.P. Beech never reached the Huron
Mts. (No. 20, Brubaker, 1975) or Vilas Co., Wisconsin (No. 21, Webb,
1974a). Dates of beech migration in lower Michigan indicate that the
species had reached the center of the peninsula (No. 8, Ahearn, unpublished; No. 9, Gilliam, et al., 1967) by 7,100 to 7,900 B.P. and
arrived at Wintergreen Lake (No. 11, R. Bailey, personal communication) by 5,700 B.P. The date of 5,100 B.P. published by Zumberge
and Potzger (1956) at South Haven (No. 10) is generally considered
to be inaccurate and too young. The diagram (Kapp et al., 1969)
from Beaver Island (No. 7) indicates Fagus arrival by about 6,000
B.P. In the southeastern corner of the state (Kapp and Kneller,
1962, No. 12), Fagus was established at Milan before 4,100 B.P. and
at nearby Frains Lake by 7,100 B.P. (Kerfoot, 1974).

   Tsuga canadensis (Eastern hemlock) pollen is equally as distinctive and even less widely disseminated from its point of origin than
is the pollen of Fagus. It's hypothetical postglacial migrational
pattern can also be postulated (Fig. 10); the same localities are
shown as for Fagus (Fig. 9). As with beech, hemlock seems to have

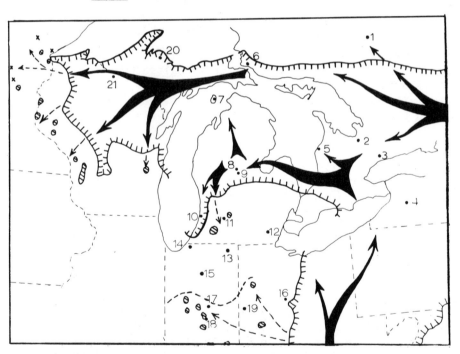

Figure 10. Postulated postglacial migrational pattern for Tsuga
canadensis (Eastern hemlock). Hachured line is limit of present
distribution (according to Fowells, 1965), except for isolated
populations which are shown in approximate locations. (See text
for site locations and references.)

entered the Great Lakes region from the south after deglaciation, probably across the drained Lake Erie basin. In contrast with Fagus, however, there is a strong likelihood that a second migrational pathway existed from eastern Pennsylvania (Martin, 1958: Tsuga present at about 10,000 B.P.) and New England (M. Davis, 1969: present at 9,000 - 10,000 B.P. at Rogers Lake, Connecticut and at 8,000 - 9,000 B.P. at Moulton Pond, Maine: R. Davis et al., 1975).

Tsuga may have entered western Ohio and central Indiana at an early date when the prevailing environment was cool and moist; in those areas it is now restricted to scattered relict stands. Forsyth (1971) cites hemlock's preference for limy soils to explain its modern distributional limits. Whether the cause be primarily climatic (related to the prairie peninsula) or edaphic, Tsuga is virtually absent from all pollen diagrams of Indiana and western Ohio (13-19), except at Silver Lake (No. 16) where it enters the record at about 10,000 B.P. and is relatively abundant. The estimated dates of Tsuga appearance at more northerly sites are: No. 2 - 10,200 B.P., No. 3 - 9,750 B.P., No. 4 - 8,500 B.P., No. 5 - 7,600 B.P., No. 6 - 6,400 B.P., and No. 1 - 7,300 B.P. As with Fagus, Tsuga seems to have entered both upper and lower Michigan from the east arriving in the central lower peninsula (Nos. 8, 9) at 7,500 - 7,900 B.P. and South Haven (No. 10) by 5,100 B.P.; it was not recorded at Wintergreen Lake (No. 11) nor was it present at Milan (No. 12) at 4,100 B.P. A definitive date from Barney Lake, Beaver Island (No. 7) places the arrival of hemlock at about 7,200 B.P. Brubaker (1975) records Tsuga in the Huron Mountains area (No. 20) by 6,000 B.P. and Webb's (1974) diagram from St. Mary's Lake (No. 21) provides a definitive date of entry there of 3,650 B.P. Either the range of Tsuga had extended as far as eastern Minnesota during recent Holocene time, only to retreat leaving relict outlying populations, or it is becoming established on these sites and still extending its natural range westward.

These postulated migrational patterns for Fagus and Tsuga require detailed testing by means of further pollen analyses at key points and by more definitive radiocarbon dating.

PROGNOSIS

Recent developments have begun to increase the sophistication of interpretation of the pollen data which have accumulated so rapdily over the past 40 years. Calculations of the absolute pollen frequency (usually by addition of known quantities of exotic pollen) and calibration of the absolute pollen influx into lake sediments (M. David, et al., 1973) have eliminated some of the uncertainties of interpretation caused by local overabundance of a certain pollen type or by irregularities in the rate of sedimentation.

Systematic studies of modern pollen rain over wide geographical areas of northeastern North America are beginning to allow statistical correlations of fossil pollen spectra with their closest modern vegetational analogues (Ogden, 1969; Webb, 1974a; 1974b; R. Davis and Webb, 1975; Brubaker, 1975).

Detailed pollen and plant macrofossil studies at key localities, both in the Great Lakes region and beyond its boundaries where northern plant communities found refugia during the full-glacial, are required for a more complete understanding of late Pleistocene geobotany. Earlier studies, coupled with these required to close the gaps identified in this paper, will provide data for applying the newly developed statistical models. The next decade should allow a rather sophisticated reconstruction of the late-glacial and postglacial vegetation of the Great Lakes region. The greatest challenge to geobotanists will be to push these studies back into pre-Wisconsinan time, a surface that has barely been scratched.

LITERATURE CITED

AHEARN, P. J. 1976. Late-glacial and postglacial pollen record from Demont Lake, Isabella County, Michigan. Unpublished senior thesis, Alma College, 17 pp.

BENNINGHOFF, W. S., 1964. The Prairie Peninsula as a filter barrier to postglacial plant migration. Indiana Acad. Sci. Proc., 72:116-124.

BERTI, A. A. 1975. Paleobotany of Wisconsinan interstadials, eastern Great Lakes region, North America. Quat. Res. 5:591-619.

BRAUN, E. L. 1950. Deciduous forests of eastern North America. Blakiston Co., Philadelphia, 596 pp.

BROWN, C. A. 1938. The flora of Pleistocene deposits in the Western Florida Parishes, West Feliciana Parish, and East Baton Rouge Parish, Louisiana. Louisiana Dept. Conserv. Geol. Bull. 12.

BRUBAKER, L. 1975. Postglacial forest patterns associated with till and outwash in north central upper Michigan. Quat. Res. 5:499-527.

COLEMAN, A. P. 1933. The Pleistocene of the Toronto region. Ont. Dept. Mines Ann. Rept. 41, part 7.

CUSHING, E. J. 1965. Problems in the Quaternary phytogeography of the Great Lakes region, p. 403-416. In: H. E. Wright, Jr. and D. G. Frey, Eds. The Quaternary of the United States. Princeton Univ. Press.

CUSHING, E. J. 1967. Late-Wisconsin pollen stratigraphy and the glacial sequence in Minnesota, p. 59-88. In: E. J. Cushing and H. E. Wright, Jr., Eds. Quaternary Paleoecology. Yale Univ. Press

DARWIN, C. 1859. The Origin of Species by Means of Natural Selection. Chapter XII, section entitled "Dispersal during the Glacial Period," (various editions).

DAVIS, J. H., JR. 1946. The peat deposits of Florida, their occurence, development, and uses. Fla. Geol. Surv. Geol. Bull., 30:1-246.

DAVIS, MARGARET B. 1965. Phytogeography and palynology of Northeastern United States, p. 377-401. In: H. E. Wright, Jr. and D. G. Frey, Eds. The Quaternary of the United States. Princeton Univ. Press.

DAVIS, M. B. 1969. Climatic changes in southern Connecticut recorded by pollen deposition at Rogers Lake. Ecology 50:409-422.

DAVIS, M. B., L. B. BRUBAKER and T. WEBB III. 1973. Calibration of absolute pollen influx, p. 9-25. In: H.J.B. Birks and R. G. West, Eds. Quaternary Plant Ecology. Blackwell Sci. Public., London.

DAVIS, R. B., T. E. BRADSTREET, R. STUCKENRATH, and H. W. BORNS. 1975. Vegetation and associated environments during the past 14,000 years near Moulton Pond, Maine. Quat. Res. 5:435-465.

DAVIS, R. B. and T. WEBB III. 1975. The contemporary distribution of pollen in eastern North America: a comparison with the vegetation. Quat. Res. 5:395-434.

DEEVEY, E. S., JR. 1949. Biogeography of the Pleistocene. Part I. Europe and North America. Bull. Geol. Soc. Amer. 60:1315-1416.

DEEVEY, E. S. 1957. Radiocarbon-dated pollen sequences in eastern North America. Veroff. Geobot. Instit. Rübel 34:30-37.

DELCOURT, H. R. and P. A.   1974.   Primeval Magnolia-Holly-Beech
    Climax in Louisiana.   Ecology 55:638-644.

DELCOURT, H. R. and P. A.   1975.   The Blufflands: Pleistocene
    pathway into the Tunica Hills.   Amer. Midl. Nat. 94:385-400.

DELCOURT, P. A. and H. R.   (In press) The Tunica Hills,
    Louisiana-Mississippi: late-glacial locality for spruce
    and deciduous forest species.   Quat. Res. (in press).

DORF, E.   1959.   Climatic changes of the past and present.
    Contrib. Mus. Paleontol. Univ. Mich. 13(8):181-210.

DREIMANIS, A.   1973.   Mid-Wisconsin of the eastern Great
    Lakes and St. Lawrence region, North America.
    Eiszeitalter und Gegenwart 23/24:377-379.

DREIMANIS, A., J. TERASMAE, and G. D. McKENZIE.   The Port
    Talbot interstade of the Wisconsin glaciation.
    Can. Jour. Eart Sci. 3:305-325.

ENGELHARDT, D. W.   1962.   A palynological study of postglacial
    and interglacial deposits in Indiana.   unpubl. Ph.D.
    thesis.   Indiana Univ.

ENGELHARDT, D. W.   1965.   A late-glacial -- postglacial
    pollen chronology for Indiana.   Amer. Jour. Sci.
    263:410-415.

FORSYTH, J.   1971.   Linking geology and botany--a new
    approach.   Explorer 13:19-25.

FORSYTH, J.   1973.   Late-glacial and postglacial history
    of western Lake Erie.   The Compass of Sigma Gamma
    Epsilon 51:16-26.

FOWELLS, H. A.   1965.   Silvics of Forest Trees of the
    United States.   U.S. Dept. Agri., Handbk No. 271,
    762 pp.

GILLIAM, J. A., R. O. KAPP, and R. D. BOGUE.   1967.   A
    post-Wisconsin pollen sequence from Vestaburg bog,
    Montcalm County, Michigan.   Pap. Mich. Acad. Sci.,
    Arts, Letters 52:3-17.

GOODING, A. M.   1966.   The Kansan glaciation in southeastern
    Indiana.   Ohio Jour. Sci. 66:426-433.

GOODING, A. M., and J. G. OGDEN, III. 1965. A radiocarbon dated pollen sequence from the Wells Mastodon Site near Rochester, Indiana.

GRAHAM, A. and C. HEIMSCH. 1960. Pollen studies of some Texas peat deposits. Ecology, 41:751-763.

GRÜGER, E. 1970. The development of the vegetation of southern Illinois since late Illinoian time (preliminary report). Rev. Geogr. Physique et de Geol. Dynamique 12:143-148.

GRÜGER, E. 1972a. Late Quaternary vegetation development in south-central Illinois. Quat. Res. 2:217-231.

GRÜGER, E. 1972b. Pollen and seed studies of Wisconsinan vegetation in Illinois, U.S.A. Bull. Geol. Soc. Amer. 83:2715-2734.

GRÜGER, J. 1973. Studies on the Late Quaternary vegetation history of northeastern Kansas. Bull. Geol. Soc. Amer. 84:239-250.

HORR, W. H. 1955. A pollen profile study of the Muscotah marsh. Univ. Kans. Sci. Bull. 37:143-149.

JUST, THEODOR. 1959. Postglacial vegetation of the north-central United States, a review: J. Geol. 67:228-238.

KAPP, R. O., S. BUSHOUSE and B. FOSTER. 1969. A contribution to the geology and forest history of Beaver Island, Michigan. Proc. 12th Conf. Great Lakes Res. 1969:225-236.

KAPP, R. O., and A. M. GOODING. 1964a. A radiocarbon-dated pollen profile from Sunbeam Prairie Bog, Darke Co., Ohio. Amer. J. Sci. 262:259-266.

KAPP, R. O. and A. M. GOODING. 1964b. Pleistocene vegetational studies in the Whitewater Basin, southeastern Indiana. J. Geol. 72:307-326.

KAPP, R. O. and A. M. GOODING. 1974. Stratigraphy and pollen analysis of Yarmouthian interglacial deposits in south-eastern Indiana. Ohio Jour. Sci. 74:226-238.

KAPP, R. O., and W. A. KNELLER. 1962. A buried biotic assemblage from an old Saline River terrace at Milan, Michigan. Pap. Mich. Acad. of Sci.,Arts and Letters 47:135-145.

KARROW, P. F., T. W. ANDERSON, A. H. CLARKE, L. D. DELORME and
    M. R. SCREENIVASA. 1975. Stratigraphy, paleontology, and
    age of Lake Algonquin sediments in southwestern Ontario,
    Canada. Quat. Res. 5:49-87.

KERFOOT, W. C. 1974. Net accumulation rates and the history of
    Cladoceran communities. Ecology 55:51-61.

KING, J. E. 1973. Late Pleistocene palynology and biogeography
    of the western Missouri Ozarks. Ecol. Monogr. 43 (4): 539-565.

MARTIN, P. S. 1958a. Pleistocene ecology and biogeography of
    North America, p. 375-420. In: Zoogeography, Amerc. Assoc.
    Adv. of Science.

MARTIN, P. S. 1958b. Taiga-tundra and the Full-Glacial per. in
    Chester County, Pennsylvania. Amer. J. Sci. 256:470-502.

McANDREWS, J. H. 1973. Pollen analysis of the sediments of the
    Great Lakes of North America, p. 76-80. In: Palynology,
    Holocene, and Marine Palynology, Proc. III Int. Paly.
    Conf. Publishing House Nauka, Moscow.

MEHRINGER, P. J., J. E. KING AND E. H. LINDSAY. 1970. A
    record of Wisconsin-age vegetation and fauna from the
    Ozarks of western Missouri, p. 173-183. In: W. Dort,
    Jr. and J. K. Jones, Jr., Eds. Pleistocene and Recent
    Environments of the central Great Plains. Univ. Kans.
    Press, Lawrence, 433 pp.

MEHRING, P. J., C. E. SCHWEGER, W. R. WOOD, and B. R. McMILLAN.
    1968. Late-Pleistocene boreal forest in the western Ozark
    highlands. Ecology 49:568-9.

MILLER, N.G. 1973a. Pollen analysis of deeply buried Quaternary
    sediments from southern Michigan. Amer. Midl. Nat. 89:217-223.

MILLER, N. G. 1973b. Late-glacial and postglacial vegetation
    change in southwestern New York State. New York State Mus.
    and Sci. Serv., Bull. 420, 102 pp.

MILLER, N. G. and W. S. BENNINGHOFF. 1969. Plant fossils from
    a Cary-Port Huron interstade deposit and their paleoecological
    interpretation. Geol. Soc. Amer., Spec. Pap. 123:225-248.

OGDEN, J. G., III. 1965. Pleistocene pollen records from eastern North America, Botan. Rev., 31:481-504.

OGDEN, J. G., III. 1966. Forest history of Ohio. Radiocarbon dates and pollen stratigraphy of Silver Lake, Logan County, Ohio. Ohio J. Sci. 66:387-400.

OGDEN, J. G., III. 1967. Radiocarbon and pollen evidence for a sudden change in climate in the Great Lakes region approximately 10,000 years ago, p. 117-127. In: Cushing E.J., and H. E. Wright, Jr., Eds. Quaternary Paleoecology. Yale Univ. Press.

OGDEN, J. G. III. 1969. Correlation of contemporary and Late Pleistocene pollen records in the reconstruction of postglacial environments in northeastern North America. Mitt. Inter. Verein. Limnol. 17:64-77.

POTZGER, J. E. 1945. The Pine Barrens of New Jersey, a refugium during Pleistocene times. Butler Univ. Bot. Stud. 7:1-15.

POTZGER, J. E. 1946. Phytosociology of the primeval forest in central-northern Wisconsin and upper Michigan and a brief postglacial history of the lake forest formation. Ecol. Monogr. 16:211-250.

POTZGER, J. E. 1948. A pollen study in the tension zone of lower Michigan, Butler Univ. Bot. Stud. 8:161-177.

POTZGER, J. E. 1951. The fossil record near the glacial border. Ohio J. Sci. 51:126-133.

POTZGER, J. E. 1952. What can be inferred from the pollen profiles of bogs in the New Jersey Pine Barrens. Bartonia 26:20-27.

POTZGER, J. E. 1953. Nineteen bogs from southern Quebec. Can. J. Bot. 31:383-401.

POTZGER, J. E., and A. COURTEMANCHE. 1956a. Pollen study in the Gatineau Valley, Quebec. Butler Univ. Bot. Stud. 8:12-23.

POTZGER, J. E., and A. COURTEMANCHE. 1956b. A series of bogs across Quebec from the St. Lawrence Valley to James Bay. Can. J. Bot. 34:473-500.

POTZGER, and J. H. OTTO. 1943. Postglacial forest succession in northern New Jersey as shown by pollen records from five bogs. Amer. J. Bot. 30:83-87.

POTZGER, J. E. and B. C. THARP. 1954. Pollen study of two bogs in Texas. Ecology 35:462-466.

SEARS, P. B. 1930. A record of postglacial climate in northern Ohio. Ohio J. Sci. 30:205-217.

SEARS, P. B. 1931. Pollen analysis of Mud Lake Bog in Ohio. Ecology 12:650-655.

SEARS, P. B. 1935a. Glacial and postglacial vegetation. Bot. Rev. 1:37-51.

SEARS, P. B. 1935b. Types of North American profiles. Ecology 16:488-499.

SEARS, P. B. 1940. Postglacial vegetation in the Erie-Ohio area. Ohio J. Sci. 41:225-234.

SEARS, P. B. 1941. A submerged migration route. Science 94:301.

SEARS, P. B. 1942a. Forest sequences in the north central states. Bot. Gaz. 103:751-761.

SEARS, P. B. 1942b. Postglacial migration of five forest genera. Amer. J. Bot. 29:684-691.

SEARS, P. B. 1942c. Xerothermic theory. Bot. Rev. 8:708-736.

SEARS, P. B. 1948. Forest sequence and climatic change in northeastern North America since early Wisconsin time. Ecology 29:326-333.

SEARS, P. B. 1951a. Palynology in North America. Svensk Bot. Tidskrift. 45:241-246.

SEARS, P. B. 1951b. Pollen profiles and culture horizons in the basin of Mexico. Proc. 29th Int. Cong. Americanists, Vol. I:57-61.

SEARS, P. B. 1952. Palynology in southern North America. II. Archeological horizons in the basins of Mexico. Bull. Geol. Soc. Amer. 63:241-254.

SEARS, P. B. 1955. Palynology in southern North America.
    Bull. Geol. Soc. Amer. 66:471-530.

SEARS, P. B. 1956. San Augustin Plains--Pleistocene
    climatic changes. Science 124:537-539.

SEARS, P. B., and M. BOPP. 1960. Pollen analysis of the
    Michillinda peat seam. Ohio J. Sci. 60:149-154.

SEARS, P. B. and K. H. CLISBY. 1952. Pollen spectra
    associated with the Orleton Farms mastodon site.
    Ohio J. Sci. 52:9-10.

SHANE, L. C. 1975. Palynology and radiocarbon chronology
    of Battaglia Bog, Portage County, Ohio. Ohio Jour.
    Sci. 75:96-102.

TERASMAE, J. 1960. A palynological study of Pleistocene
    interglacial beds at Toronto, Ontario. Geol. Surv.,
    Canada, Bull. 56:23-41.

TERASMAE, J. 1967. Postglacial chronology and forest
    history in the northern Lake Huron and Lake Superior
    regions, p. 45-58. In: E. J. Cushing and H. E.
    Wright, Jr., Eds. Quaternary Paleoecology, Yale Univ.
    Press.

TURNER, C. and R. G. WEST. 1968. The subdivision and
    zonation of interglacial periods. Eiszeitalter und
    Gegenwart 19:93-101.

VINCENT, J. 1973. A palynological study for the Little
    Clay Belt, northwestern Quebec. Naturaliste Can.
    100:59-70.

VON POST, L. 1916. Om skogstradpollen i sydsvenska torr-
    mosslager-földer. Geol. for. Stockholm forhdl. 38:384.

VOSS, J. 1933. Pleistocene forest of central Illinois.
    Bot. Gaz. 94:808-814.

VOSS, J. 1934. Postglacial migration of forests in Illinois,
    Wisconsin, and Minnesota. Bot. Gaz. 96:3-43.

WATTS, W. A. 1970. The full-glacial vegetation of northwestern
    Georgia. Ecology 51:18-33.

WATTS, W. A. 1975. Vegetation record for the last 20,000 years from a small marsh on Lookout Mountain, northwestern Georgia. Bull. Geol. Soc. Amer. 86:287-291.

WEBB, T. III. 1974a. A vegetational history from northern Wisconsin: evidence from modern and fossil pollen. Amer. Midl. Nat. 92:12-34.

WEBB, T. III. 1974b. Corresponding patterns of pollen and vegetation in Lower Michigan: A comparison of quantitative data. Ecology 55:17-28.

WEST, R. G. 1961. Late and postglacial vegetational history in Wisconsin, particularly changes associated with the Valders readvance. Amer. J. Sci. 259:766-783.

WHITEHEAD, D. R. 1967. Studies of full glacial vegetation and climate in southeastern United States, p. 237-248. In: E. J. Cushing and H. E. Wright, Jr., Eds. Quaternary Paleoecology. Yale Univ. Press, New Haven, Connecticut, 433 pp.

WHITEHEAD, D. R. 1973. Late-Wisconsin vegetational changes in unglaciated eastern North America. Quat. Res. 3:621-631.

WRIGHT, H. E., JR. 1964. Aspects of the early postglacial forest succession in the Great Lakes region. Ecology 45:439-448.

WRIGHT, H. E., JR. 1968a. The roles of pine and spruce in the forest history of Minnesota and adjacent areas. Ecology 49:937.

WRIGHT, H. E., JR. 1968b. History of the prairie peninsula, p. 78-88. In: The Quaternary of Illinois, Univ. Ill. Coll. Agric., Spec. Publ. 14, 179 pp.

WRIGHT, H. E., JR. 1971. Late Quaternary vegetational history of North America, p. 425-464. In: K. Turekian, Ed. The Late Cenozoic Glacial Ages. Yale Univ. Press.

WRIGHT, H. E., JR. 1972. Interglacial and postglacial climates: the pollen record. Quat. Res. 2:274-282.

ZUMBERGE, J. H., and J. E. POTZGER. 1956. Late Wisconsin chronology of the Lake Michigan basin correlated with pollen studies. Bull. Geol. Soc. Amer. 67:271-288.

MEMORIAL TRIBUTE - ANSEL M. GOODING

This paper is dedicated, with deep respect, as a memorial to Ansel M. Gooding.

Ansel M. Gooding (1924-1976), Professor of Geology at Earlham College, Richmond, Indiana, had planned to attend the February 21, 1976 Geobotany Conference at Bowling Green State University but was hospitalized following a heart attack during the prior week. He was regaining strength during the spring and had resumed limited teaching activities from his home when a relapse claimed his life on March 24.

An especially perceptive field geologist, Gooding provided the definitive interpretation of the glacial geology of southeastern Indiana and adjacent Ohio. Interdisciplinary in his approach, he directed studies of soils and paleosols and continuously discovered pollen-rich Pleistocene deposits. Interstadial and interglacial deposits are common in his research region near the southern limits of Wisconsinan and Illinoian glaciation. It has been my privilege to study the pollen and macrofossils in many of these sediments, beginning in 1959. Totally dedicated to excellence in both undergraduate teaching and in research, Gooding inspired both student and professional colleagues to extend the understanding of Pleistocene geology and paleoecology. The geobotanically-rich region in which Gooding was working will doubtless yield many more fossiliferous deposits; Gooding's research and personal commitments will inspire others to continue geobotanical studies in the southern Great Lakes region.

LIMITING FACTORS IN PALEOENVIRONMENTAL RECONSTRUCTION

J. Gordon Ogden, III

Department of Biology, Dalhousie University

Halifax, Nova Scotia, CANADA

ABSTRACT

It is a basic tenet of paleoecologists that the data set, consisting principally of pollen, spores, diatoms, or other microfossils
and the enclosing sediment, is better than anything yet done with it.
Distributional problems include differential pollen production, dispersal, sedimentation, and preservation. Changes in dominant air
mass patterns, whether seasonal or in response to climatic trends,
alter source pollen frequency independently of vegetational sources
near depositional sites. Approximately 50% of the pollen at the
center of a small (ca 1 km diam.) pond comes from within 7 km of
the site, and from less than 300m of the source in a bog. With more
than 300 published pollen sequences and nearly 500 contemporary
pollen records available in North America since the 1930s, the sampling interval, even if uniformly spaced, is ca 10^4 km^2 per sample
site. At this spacing, approximately 11 stations at 110 km intervals
would be available to describe the vegetational diversity of Ohio.

External influences upon the data set include site-specific
variables such as topography, exposure, soil type and geology, and
trophic status of the depositional site. Increasing attention to
sampling and quantitative preparation techniques, inclusing close
interval stratigraphic control by radiocarbon dates, is improving
the precision of reconstruction of the stratigraphic record. Prolif-
eration of large computers and sophisticated statistical procedures
promise isolation and identification of quantitative changes in
important climatic variables.

Despite the necessity for increased sample coverage, a growing responsibility to fully utilize existing records and preserve existing and potential sites for future more sophisticated investigations must be recognized.

INTRODUCTION

The inferences of paleoecology rest upon two fundamental premises, the implications and limitations of which must be considered in any evaluation of fossil evidence. The first is that organisms in the past had environmental requirements similar to those of their modern counterparts. The second assumption is that a given biotic population is in equilibrium with its environment, and that the population will show no major change unless there is alteration of the environment.

The validity of paleoecologic inferences decrease with increasing geologic age. For this reason, the techniques of paleoecology are restricted, except in very broad terms, to more recent geologic time. In particular, the floras of the Cenozoic are susceptible to this treatment, since most of the plant genera from Upper Cretaceous and later strata are clearly referable to modern genera. Confidence in the equivalence of vegetational units increases as later Pleistocene units are considered. Available evidence indicates close correspondence of vegetational associations in Wisconsinan and younger deposits.

Since 1930, the microfossil stratigraphy, especially of pollen, has been reported from more than 300 sites in northeastern North America. More than one-half of these studies were based on tree pollen counts alone, which makes it difficult to infer the treeless conditions characteristic of tundra environments. The advent of radiocarbon dating in the 1950s made possible the calibration of pollen-stratigraphy sequences, utilizing an independent geochemical chronology. Fewer than one-third of the sequences in northeastern North America have any radiocarbon determinations, and only about 25 sequences have more than five radiocarbon determinations on individual cores which permit reliable calculation of sedimentation rates.

Increasing sophistication of sample preparation and counting techniques, spearheaded by M. B. Davis and her coworkers (1963, 1965, 1967, 1967b,1968,1971,1973) have produced pollen diagrams based on absolute pollen influx rates, e.g., as pollen grains $cm^{-2} yr^{-1}$. These techniques have a principal advantage of removing the statistical restrictions of "closed universe" populations based on proportions.

The explosive development of larger and more powerful computers and software since the mid-1960s has provided the pollen analyst with tools of spectacular power. Manipulation of large data sets by power-

ful statistical treatment places an increasing responsibility upon
the investigator to be sensitive to distributional problems associated with data sets of uneven quality, e.g., with or without non-
arboreal pollen, moss polsters vs. lake, bog, or atmospheric samples,
absolute or relative pollen counts, and counting to fixed or variable
pollen sums. Preliminary efforts to reconcile pollen and vegetational
records by Ogden (1969) have been refined by Davis and Webb (1975),
Webb (1974), Bernabo and Webb (in press), and are the subject of active investigation by numerous other investigators (Andersen, 1970;
Janssen, 1966, 1967, 1970, 1973; Lichti-Federovitch and Ritchie, 1968).
Some of the more intractable problems under study include pollen production (Wright, 1952), pollen dispersion, filtration, and deposition
(Tauber, 1965, 1967; E. C. Ogden, 1964; Raynor, 1965; Janssen, 1966),
resuspension and redistribution of pollen grains in lakes (Davis, 1968),
statistical treatment of pollen data (Mosimann, 1962, 1965, 1971; Ogden,
1969; Yarranton and Ritchie, 1972; Webb and Bryson, 1972; Webb, 1974).
Increasing attention to paleoclimatology (Lamb, 1966, 1970 ; Bryson, et
al, 1967) and the application of powerful multivariate statistical tools
has enabled the use of transfer functions to contrast pollen records
and vegetation types with climatic shifts, such as weather-generating
air mass movements since deglaciation (Webb and Bryson, 1972). Among the
exciting possibilities inherent in these approaches are recognition
of changes in the length of the growing season, as well as other important climatic variables of plant growth and distribution.

Difficult problems remain, however, as Ogden (1969) pointed out,
"The use of statistical procedures can scarcely be justified unless it
is possible to derive inferences and conclusions from their use that
would not be possible otherwise. The further restriction that applic-
able statistical procedures must be free from assumptions about the distribution of the items under consideration limits the kind of statistical treatments possible...."

The first of these constraints, that of deriving inferences by
statistical procedures that could not otherwise be recognized, is being
admirably tackled by Webb (1974) in his use of principal component
analysis in contrasting pollen spectra with vegetational records in
lower Michigan; by Davis and Webb (1975), and Bernabo and Webb (in press)
in the construction of isopollen maps overlaid on vegetation maps. The
second restriction, that of appropriate statistical procedures, is
less clear. Mosimann (1962,1965) and Mosimann and Greenstreet (1971)
have provided a number of useful statistical methods for handling pollen
data, but critical tests of between and within group variances have
yet to be performed. These observations are not intended to denigrate
the excellent studies now published or in progress, but as a caution
against acceptance of inferences whose apparent rigor may be diluted
by unresolved distributional problems. It is instructive to recall
that few of the conclusions of earlier workers (for reviews, see
Sears, 1935 and Ogden, 1965) have been discarded. The generalization that the data set is better than anything yet done with it, still
holds true.

NATURE AND CHARACTERISTICS OF THE DATA SET

The area of North America considered here lies between 55° and 95° W Long and 40° and 65° N Lat, covering an area of approximately 3.7×10^6 km^2. Environmental variability ranges from permanently frozen ground (N of ca 56° N Lat) to about 3 months of shallow ground frost (ca 40° N Lat). Growing seasons range from less than 30 frost free days in the north to more than 150 days in the southern portions of the range. The consequent matrix of climatic and edaphic conditions have produced a vegetational mosaic whose diversity is exceeded only by tropical plant associations. Communities range from treeless high arctic tundra through conifer, conifer-hardwood, and complex deciduous forest associations whose components may not have been displaced since Tertiary time.

The primary tool to reconstruct displacement and migration of this vast vegetational complex is the contemporary pollen record of these associations. It must be remembered that pollen grains are plants, the gametophyte generation (for spermatophytes) of the plant's life history. The sporophytes of major vegetational units are committed to survival or extinction in one place as soon as the radicle emerges. Climatic and edaphic factors, tempered by topography, exposure, fire or biotic controls (grazing, parasitism, disease), determine the success or failure of individual elements, which in the aggregate determine the stability or direction of migration of a vegetational association.

Isochrones of Wisconsinan Ice Retreat

Figure 1 shows approximate ice-front retreatal positions during Wisconsinan deglaciation and is adapted from Prest (1969) and Bryson (1969). Although agreement is close, an error of ca 1000 years may be assumed for specific ice front positions. Black dots show pollen-stratigraphic sequences which have 5 or more radiocarbon control determinations. Several sequences have 7 or more dates. Open circles indicate sequences with 3 or fewer radiocarbon determinations. Least squares power curve regressions of the form $\hat{x}(\text{age}) = ay^b$ (depth in cm) provide consistent estimates ($r^2 > .97$) of sedimentation rates.

The scale of the dots in Fig. 1 indicate an area coverage of ca 100 km^2. According to Tauber (1967) more than 50% of the pollen recovered from a small (ca 1 km diam) lake would come from the area covered by a dot. The most complete area coverage to date is from Michigan, due largely to the work of M. B. Davis, et al (1973) and Webb (1974). Extensive sampling and collation of existing data by R. B. Davis and Webb (1975) has provided nearly 500 surface samples from eastern and central North America. Even if evenly distributed, this coverage would provide a sampling interval of ca 110 km, a grid

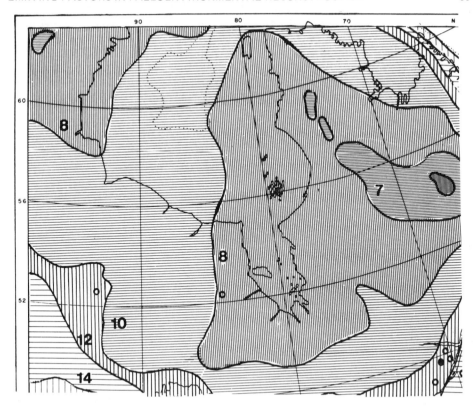

Figure 1. Isochrones of late Wisconsinan ice retreat. Solid circles indicate pollen sequences controlled by 5 or more radiocarbon dates. Open circles indicate sequences with 3 or fewer radiocarbon dates. After Prest (1969) and Bryson (1969).

inadequate to reveal details of the vegetational mosaic of the northeast. Because the sampling intervals are not uniform, it follows that vast areas of pollen sources have yet to be adequately sampled. In spite of these reservations, Davis and Webb (1975) were able to conclude,"This study (sic) shows a strong correspondence between the pollen and vegetation on a continental scale. Each major vegetational region is characterized by a distinctive pollen assemblage, and the distribution of major pollen types show consistent mappable patterns." Their study further indicates that 0 isopoll limits correspond closely with the range limits of parent plants, a conclusion reached earlier by Ogden (1969) on less objective grounds. Ogden also concluded that within lake variability of pollen samples was substantially less (for central lake-basin sites) than differences between major vegetation types in the eastern Great Lakes region.

Terminal emplacement of the Late Wisconsinan ice advance was not everywhere synchronous, but achieved maximum extent roughly 16,000 to 18,000 years ago. In general, western lobes of major ice segments were emplaced and began to retreat prior to maximum advance and subsequent retreat of eastern lobes. Thus, in Ohio, the Miami lobe reached its maximum extent about 18,000 years ago, and the eastern, or Scioto lobe reached its maximum about 16,000 years ago (Dreimanis and Goldthwait, 1973). Similarly, in southeastern New England, the Naragansett lobe (eastern) overran the eastern edge of the Point Judith (western) lobe. On Cape Cod, the Buzzards Bay ice (western) began to retreat before the emplacement of the Sandwich (eastern) ice as shown by Sandwich drift deposited inside the Buzzards Bay morainal system (Mather, Goldthwait, and Thiesemeyer, 1942). In any event, ice retreat became general shortly after 16,000 years ago, and by 14,000 years ago, a broad belt of recently deglaciated landscape existed throughout the entire region.

Most of the radiocarbon dated pollen sequences shown in Fig. 1 were ice-free by 14,000 years ago, yet few of the sequences were accumulating pollen records prior to about 12,000 years ago. Notable exceptions include West Okoboji Lake site (Bender and Bryson, in press), Kirchner Marsh, Minnesota, Lawrence Lake, Michigan, Pretty Lake, Indiana, Torren's Bog, Ohio, and Moulton Pond, Maine (Ogden, in press). Although other undated sequences can probably be included, it is apparent that 1000-2000 years have elapsed between deglaciation and the accumulation of sedimentary records in most eastern North America sites. It is therefore not surprising that evidence of tundra environments has remained elusive. Ogden (1965) reviewed evidence for high arctic environments in eastern North America, noting a paucity of evidence from the midwest. Bones of high arctic animals from the New Paris sinkhole site in western Pennsylvania (Guilday,1963, Guilday and Doutt, 1961, Guilday and Bender, 1960) remain the most convincing arguments for high arctic environments in the midwest. More recently, Ogden and Hay (1969) described anarctic pollen flora associated with a radiocarbon date of 19,730 ± 475 y.B.P. (OWU-257) from southwestern Ohio. In the northeast, apart from occasional finds of Armeria sibirica and associated pollen (Ogden, 1965), the most convincing evidence has come from Moulton Pond, Maine (R. B. Davis, et al, 1975) where pollen influx diagrams reveal virtually treeless conditions until less than 10,000 years ago. On less certain grounds, similar tundra conditions are inferred for the Lac Louis region, Quebec (Vincent, 1973) and for the Lac Mimi area (Richard and Poulin, 1975).

Consideration of the retreatal isochrones implies major downwasting and melt-back between 14,000 and 12,000 years ago. Establishment of St. Lawrence drainage was not complete until slightly less than 12,000 years ago. Since a major portion of the Great Lakes drainage system was therefore forced through a Hudson River outlet, torrential meltwaters must have created a considerable barrier to northeasterly migrations of early man. The Duchess Quarry Cave site in Orange County, New York, located just west of the Hudson River drainage channels includes a radiocarbon date of 12,540 $\pm$ 370 y.B.P. (I-4137, Funk, et al, 1970), with broken caribou bones indicating human activity. The surrounding region has yielded numerous remains of mammoth, mastodon, giant moose-deer, Ground Sloth, American Bison, and Caribou (Fisher and Reilly, 1969).

Although the Bull Brook site in eastern Massachusetts is inferred to be older than the radiocarbon dates associated with it (Byers, 1975), the Debert site near Truro, Nova Scotia remains an uncontestable argument for the presence of early man in the northeast. Firmly dated by Stuckenrath (1966) at ca 10,500 years, B. P., Byers, 1975) infers that this site, near the head of Minas Basin in the Bay of Fundy, was occupied for a period of 215 to 900 years.

By 10,000 years ago, ice had retreated north of the Great Lakes region and the St. Lawrence, and all of the residual ice caps in New England and Maritime Canada had disappeared. The locations and activities of man during the period 10,000 to 6000 y. B. P. remain disappointingly obscure (Byers, 1975). Recent work by Tuck in Labrador and coastal Newfoundland (McGhee and Tuck, 1975, Tuck, in press) reinforces the impression that early man in the northeast confined his activities to the immediate coastal zone. Massive isostatic and eustatic adjustments of the coastline have resulted in submergence of many important sites. At the same time, in Maine (Sanger, et al, in press) and in Newfoundland/Labrador (Tuck, in press), large areas of coastline have emerged due to isostatic rebound. Resolution of many of the problems of early man in the northeast can be anticipated as radiocarbon control is extended to these areas.

Separation of the Keewatin and Labradorean ice massifs by 8000 years ago permitted the extension of the Tyrell Sea into the lowlands formerly occupied by Lakes Ojibway-Barlow and Agassiz west of James Bay. Torrential meltwaters, presumable heavily silt-laden, restricted the productivity of the freshly exposed area.

By 6000 y. B. P., all of the northeast was ice-free, the last ice disappearing from the highlands of the Torngat Mountains of Labrador shortly after 6500 y. B. P. (Fig. 1). Modern drainage from the Great Lakes and the St. Lawrence was established by closure of the North Bay/Ottawa River outlet of the western Great Lakes, and the development of northward drainage as the James Bay lowlands decanted into northern waters.

Continued climatic amelioration culminated in the postglacial warm-dry maximum, recognized in Ohio as the Xerothermic Interval

(Sears, 1942a), ca 4500-3500 y.B.P. Subsequent climatic deterioration resulted in locally important readvances of valley glaciers but did not result in reestablishment of major centers of ice accumulation.

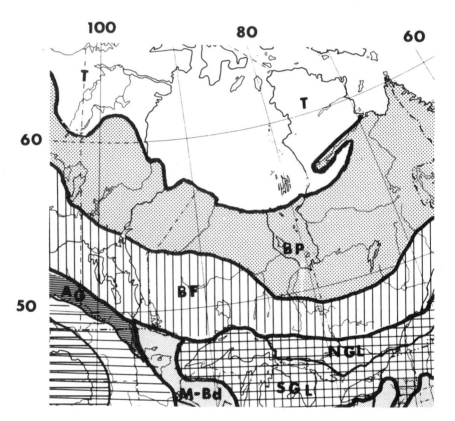

Figure 2. Major vegetational units of northeastern North America.
T = Tundra; BP = Boreal Parkland; BG = Boreal Forest; AG = Aspen-Grassland; NGL = Northern Great Lakes-St. Lawrence Forest; SGL = Southern Great Lakes-St. Lawrence Forest; SG = Short grass prairie; TG = Tall Grass prairie; M-Bd = Maple-Basswood; O-H = Oak-Hickory; B-M = Beech-Maple; O-C = Oak-Chestnut; O-P = Oak-Pine; MM = Mixed Mesophytic. (After Ogden, 1969).

Major Vegetational Units

Since the principal tool for environmental reconstruction continues to be pollen analysis, it is appropriate to consider the "grain size" of the vegetational units recognizable in pollen diagrams. Early attempts to link vegetational units with pollen sequences resulted in zonation of pollen diagrams into "Spruce-Fir zone," or "Oak-Hickory," recognizing dominant pollen types as representing identifiable vegetational units. These units are recognizable over wide areas of northeastern North America and provide reassuring consistency in postglacial records.

Numerous approaches to refinement of these patterns are recorded. Ogden (1969) utilized a Spearman Rank Correlation procedure to compare ca 200 surface pollen spectra from different vegetational regions with core data from lakes in Ohio and Indiana. Recognizable vegetational units (Spearman $r_s > .800$) are shown in Figure 2. More detailed reconstructions are described in Janssen (1970), Webb (1974), Davis and Webb (1975), and Bernabo and Webb (in press).

It is apparent that continued refinement will require more than doubling of the number of existing surface pollen samples. Ogden (in press) shows the existing number of contemporary sample points at a scale of ca 100 km² for each sample site. Large areas remain to be sampled for more accurate vegetational reconstruction. Only in Michigan, where Webb (1974) has contrasted contemporary pollen samples with existing vegetation, is the sampling density adequate for detailed reconstruction.

The concept of a distinctive pollen "signature" characteristic of a mappable vegetational unit underlies the studies of Ogden, Webb, and Davis referred to above. This approach enabled Ogden (1969) to infer the presence of Maple-Basswood forest (M-Bd in Fig. 2) several hundred kilometers southeast of its present range ca 6000 years ago. The result is significant since both Maple and Basswood are insect pollinated and provide little pollen evidence of their dominance in the forest. Nevertheless, their presence in the regional vegetation distorts the local pollen rain in a characteristic fashion, resulting in a statistically significant "signature" despite the low incidence of maple or basswood pollen in the profile.

The ability of computers to "see" patterns in multivariate data sets is probably the strongest justification for their application to paleoecological problems.

TRENDS AND PROSPECTS

Paleoecologists have generally been reluctant to adopt quantitative approaches to environmental reconstruction. The availability of

large computers, capable of manipulating massive data sets, together
with sophisticated statistical and library programs, provide an
opportunity for far greater detail in the reconstruction of paleoenvironments.

Despite the necessity for increased sampling density, both for
cores and for contemporary pollen records, paleoecologists must become
more sensitive to an increasing responsibility to fully utilize existing records, and to protect and preserve existing and potential sites
for future more sophisticated study. Where possible, sediment coring
should be carried out over ice, to protect stratigraphic records from
disturbance by mooring anchors and equipment. As a case in point, Silver Lake, Ohio (Ogden, 1966, 1969) has only a small remaining undisturbed area in the central basin (for which the author admits responsibility) as a result of more than a dozen short and long cores taken
in the course of various investigations.

Broader use of geochemical tools, such as close interval radiocarbon dating (5 or more determinations per core), oxygen isotope and
stable carbon ($^{13}C/^{12}C$) isotope ratios, tritium and/or lead-204 dating
of recent sediment sequences, is imperative.

Because the impact of man is so pervasive, the influence of man's
activities is fully as much a part of the story as the climatically
induced changes usually considered. With the capacity to drastically
alter environments comes the concommitant responsibility to recognize
future environmental change and trends. The paleoecologist is in a
unique position to assess the impact of proposed land use practices.
This responsibility defines the necessity for increased quantitative
rigor in paleoenvironmental studies. While the past is unquestionably
the key to the present, there is an equal or greater responsibility
to the future.

LITERATURE CITED

Andersen, S. Th. 1970. The relative pollen productivity and pollen
    representation of north European trees, and correction factors for
    tree pollen spectra. Danm. Geol. Undersøg. Ser. II, 96. 99p.

Bernabo, J. C. and T. Webb. (in press). Changing patterns in the Holo-
    cene pollen record of northeastern North America: A mapped summary.
    Quat. Res.

Bryson, R. A. and W. M. Wendland. 1967. Tentative climatic patterns
    for some late-glacial and postglacial episodes in central North
    America. In Life, Land and Water, W. J. Mayer-Oakes, Ed. 271-289.
    Univ. of Manitoba Press, Winnipeg.

Bryson, R. A., D. A. Barreis, and W. M. Wendland.1970. The character of late- and postglacial climatic changes. In Pleistocene
and recent environments of the central Great Plains. W. Dort
and J. K. Jones, eds. 53-74. Univ. of Kansas Press, Lawrence.

Bryson, R. A., W. M. Wendland, J. D. Ives, and J. T. Andrews. 1969.
Radiocarbon isochrones on the disintegration of the Laurentide
ice sheet. Arctic and Alpine Res. 1:1-14.

Byers, D. S. 1975. Environment and subsistence. In Environmental
change in the Maritimes. J. G. Ogden, III and M. J. Harvey, eds.
3-16. N. S. Inst. Sci. v. 27, suppl. 3, Halifax.

Davis, M. B. 1963. On the theory of pollen analysis. Amer. Jour.
Sci. 261: 897-912.

_____. 1965. A method for determining absolute pollen fre-
quency. In Handbook of paleontological techniques. B. Kummel and
D. Raup, eds. 674-686. Freeman and Co. San Francisco.

_____.1967a. Pollen deposition in lakes as measured by sedi-
ment traps. Geol. Soc. Amer. Bull. 78: 849-858.

_____. 1967b. Pollen accumulation rates at Rogers Lake, Conn-
ecticut during late- and postglacial time. Rev. of Paleobotany
and Palynology 2: 219-230.

_____. 1968. Pollen grains in lake sediments: redeposition
caused by seasonal water circulation. Science 162: 796-799.

Davis, M. B., L. B. Brubaker, and J. Beiswanger. 1971. Pollen grains
in lake sediments: pollen percentages in surface sediments from
southern Michigan. Quat. Res. 1: 450-467.

Davis, M. B., L. B. Brubaker, and T. Webb, III. 1973. Calibration of
absolute pollen influx. In Quaternary Plant Ecology. H. J. B.
Birks and R. G. West, eds. 9-25. J. Wiley, New York.

Davis, R. B., T. E. Bradstreet, R. Stuckenrath, Jr., and H. W. Borns,
Jr. 1975. Vegetation and associated environments during the past
14,000 years near Moulton Pond, Maine. Quat. Res. 5:436-465.

Davis, R. B. and T. Webb, III. 1975. The contemporary distribution
of pollen in eastern North America: a comparison with the vegetation. Quat. Res. 5: 395-434.

Fisher, D. W., and E. Reilly. 1969. New discoveries of late Pleistocene mammals in the Hudson valley. Geol. Soc. Amer. Prog.
with abstracts for 1969, Part I, meeting in Albany, N.Y., p.18.

Funk, R. E., D. W. Fisher, and E. M. Reilly, Jr. 1970. Caribou and paleo-Indian in New York state: a presumed association. Amer. J. Sci. 268: 181-186.

Guilday, J. E. 1963. Pleistocene zoogeography of the lemming, Dicrostonyx. Evolution 17: 194-197.

Guilday, J. E. and J. K. Doutt. 1961. The collared lemming, (Dicrostonyx) from the Pennsylvania Pleistocene. Proc. Biol. Soc. Washington 74: 249-250.

Guilday, J. E. and M. S. Bender. 1960. Late pleistocene records of the yellow-cheeked vole, Microtus xanthognathus (Leach). Ann. Carn. Mus. 35: 315-330.

Janssen, C. R. 1966. Recent pollen spectra from the deciduous and coniferous forests of northwestern Minnesota: a study in pollen dispersal. Ecol. 47: 804-825.

_____. 1967. A postglacial pollen diagram from a small Typha swamp in northwestern Minnesota, interpreted from pollen indicators and surface samples. Ecol. Monog. 37: 145-172.

_____. 1970. Problems in the recognition of plant communities in pollen diagrams. Vegetatio 20:187-198.

_____. 1973. Local and regional pollen deposition. In: Quaternary Plant Ecology. H. J. B. Birks and R. G. West, eds. 31-42. J. Wiley, New York.

Lamb, H. H. , R. P. W. Lewis, and A. Woodroffe. 1966. Atmospheric circulation and the main climatic variables between 8000 and 0 B. C.: Meteorological evidence. In World Climate from 8000 to 0 B. C. T. Sawyer, ed. 174-217. Royal Meteorol. Soc. London.

Lamb, H. H. and A. Woodroffe. 1970. Atmospheric circulation during the last ice age. Quat. Res. 1: 29-58.

Lichti-Federovitch, S. and J. C. Ritchie. 1968. Recent pollen assemblages from the western interior of Canada. Rev. Paleobotany and Palynology 7: 297-344.

McGhee, R. and J. A. Tuck. 1975. An archaic sequence from the Strait of Belle Isle, Labrador. National Museum of Man Mercury Series, ISSN 0316-1854, National Museums of Canada. Ottawa. 254p.

Mosimann, J. E. 1962. On the compound multinomial distribution, the multivariate β-distribution, and correlations among proportions. Biometrika 49: 65-82.

_____. 1965. Statistical methods for the pollen analyst: Multinomial and negative multinomial techniques. In Handbook of Paleontological Techniques. B. Kummel and D. Raup, eds. 636-673. Freeman, San Francisco.

Mosimann, J. E. and R. I. Greenstreet. 1971. Representation insensitive methods in paleoecological pollen studies. In Statistical Ecology, Vol. I, Spatial patterns and statistical distribution. K. Patel and E. C. Pielou, eds. 23-58. Penn State Press.

Ogden, E. C., G. S. Raynor, and J. M. Vormevik. 1964. Travels of airborne pollen. New York State Museum and Science Service, Progress Report No. 5. Albany, N.Y.

Ogden, J. G., III. 1965. Pleistocene pollen records from eastern North America. Bot. Rev. 31: 481-504.

_____. 1969. Correlation of contemporary and late Pleistocene pollen records in the reconstruction of postglacial environments in northeastern North America. Mitt. Internat. Verein. Limnol. 17: 64-77.

_____. (in press). The late Quaternary paleoenvironmental record of northeastern North America. Proc. AMERIND Conf. N.Y. Acad. Sci. 4-6 Feb. 1976. New York.

Ogden, J. G. III, and R. J. Hay. 1969. Ohio Wesleyan University natural radiocarbon measurements IV. Radiocarbon 11: 137-149.

Prest, V. K. 1969. Retreat of Wisconsin and Recent ice in North America. Map 1257A. Geol. Survey of Canada. Ottawa.

Raynor, G. S. and E. C. Ogden. 1965. Twenty-four hour dispersion of ragweed pollen from known sources. Brookhaven Natl. Lab. Report BNL 957 (T-398). Upton, N.Y.

Richard, P. and S. Poulin. 1973. Un diagramme pollinique au Mont des Eboulements, region de Charlevoix, Québec. Can. J. Earth Sci. 13: 145-156.

Sanger, D., R. Davis, R. MacKay, and H. W. Borns, Jr. (in press). Paleoenvironments and prehistoric man at the Hirundo site, Maine. Proc. AMERIND conf. N.Y. Acad. Sci. 4-6 Feb., 1976. New York.

Sears, P. B. 1935. Glacial and postglacial vegetation. Bot. Rev. 1: 37-51.

_____. 1942. Xerothermic Theory. Bot. Rev. 8: 708-736.

Tauber, H. 1965. Differential pollen dispersion and the interpretation of pollen diagrams. Danm. Geol. Undersøg. Ser. II. 89. 69p.

_____. 1967. Differential pollen dispersion and filtration. In Quaternary Paleoecology. E. J. Cushing and H. E. Wright, Jr. eds. 131-141. Yale Univ. Press. New Haven.

Tuck,J. A. (in press). Early Archaic cultures in the Strait of Belle Isle, Labrador. Proc. AMERIND Conf. N. Y. Acad. Sci. 4-6 Feb., 1976. New York.

Vincent, J-S. 1973. A palynological study for the Little Clay belt, northwestern Quebec. Naturaliste Can. 100: 59-70.

Webb, T., III. 1974. Corresponding patterns of pollen and vegetation in lower Michigan: A comparison with quantitative data. Ecol. 55: 17-28.

Webb, T. III, and R. A. Bryson. 1972. Late- and postglacial climatic change in the northern midwest USA: Quantitative estimates derived from fossil pollen spectra by multivariate statistical analysis. Quat. Res. 2: 70-115.

Wright, J. W. 1952. Pollen dispersion of some forest trees. U. S. Forest Service, Northeastern Forest Expt. Sta. Paper 46.

Yarranton, G. A. and J. C. Ritchie. 1972. Sequential correlations as an aid in placing pollen zone boundaries. Pollen et Spores XIV: 213-223.

TAXONOMIC AND STRATIGRAPHIC SIGNIFICANCE OF THE DISPERSED SPORE GENUS CALAMOSPORA

Charles W. Good

Department of Botany

Ohio State University, Lima, Ohio   45804

ABSTRACT

In situ spores are described from Pennsylvanian-age sphenopsid cones. Spores from cones referable to the plant Calamites have been shown to resemble the dispersed spore genera Elaterites, Vestispora, or Calamospora at different stages in spore ontogeny. Spores from some cones referable to the plant Sphenophyllum have ontogenetic stages resembling Vestispora and Calamospora, indicating that these two dispersed spore genera were produced by more than one type of plant. To date, features used to delimit species of dispersed Calamospora-like spores have included spore diameter, presence or absence of dark contact areas, distribution and arrangement of compression folds, minute wall ornamentation, presence of a globular inner body, separation of presumed outer and inner exine layers, and length of trilete arms. Results of the present investigation indicate that none of these features can be used to distinguish between spores produced by different cone species. The wide range of variation in these features among spores obtained from different specimens of the same cone species, and even from the same cone specimen, suggest that a clearcut division of species within the dispersed spore genus Calamospora is not possible. Caution should be used in making stratigraphic determinations based largely from dispersed spores that resemble Calamospora, Elaterites, Vestispora, Schopfites, Phyllothecotriletes, Ricaspora, and Perotriletes.

INTRODUCTION

Some stratigraphic studies have been based, at least in part, upon dispersed spore floras. In order to assist in spore classification and identification, palynologists typically assign genus and species names to each spore type which can be recognized as morphologically distinct. The use of genus and species names does not necessarily imply any natural relationships among similar spores. Nevertheless, there is an unstated assumption in this taxonomic procedure, namely that each species of dispersed spore was produced by a different plant. In sediments older than Tertiary age, it is usually only possible to identify the parent plant which produced a particular type of fossil spore by finding spores preserved in situ within fossil fructifications of the parent plant.

This report describes in situ Pennsylvanian-age spores which, if found dispersed, would be classified within the dispersed spore genus Calamospora. This genus was erected by Schopf et. al. (1944) to include spherical trilete smooth-walled sporomorphs ranging in size from 20 to several hundred micrometers in diameter. As such, Calamospora is one of the morphologically simplest spore types and is the only dispersed spore genus known to include megaspores, microspores, and isospores. Currently available literature indicates that Calamospora is found in rocks of Devonian through Triassic age (Hoffmeister et. al., 1955; Mädler 1964), although the genus is most commonly reported from the Mississippian and Pennsylvanian Systems.

In the past, Calamospora has been identified as representative of spores produced by the plant Calamites because published illustrations have shown what appear to be Calamospora-like spores within sporangia of calamitean cones (Abbott, 1968; Delevoryas, 1955; Hartung, 1933; Moore, 1946; Taylor, 1967). However, recent investigations (Good and Taylor, 1974, 1975; Good, 1975) have shown that mature spores of many, and possibly all, calamitean cones correspond to the dispersed spore genus Elaterites Wilson (1943). When found within sporangia of well preserved calamitean cones (Fig. 1), such mature spores bear three superficially attached strap-shaped structures (Figs. 4, 5) similar to the elaters of modern Equisetum spores. Immature calamitean spores with tightly coiled elaters (Fig. 3 right) resemble the dispersed spore genus Vestispora Wilson and Hoffmeister (1956). It is only when calamitean spores have shead their elaters, or when the perisporial elater layer is not preserved, that these spores show a Calamospora-like configuration (Fig. 7). Figures 3, 4, 5, and 7 illustrate Elaterites, Vestispora, and Calamospora-like spores of the same cone species, Palaeostachya decacnema.

Considering the morphological simplicity of dispersed
Calamospora specimens, a surprising number of species have been
established within this genus. Characteristics used in defining
these species have included the presence or absence of darkened
contact areas (area contagionus) between arms of the trilete,
presence of a globular body internal to the spore exine, separation of outer and inner exine layers, exine thickness, ratio of
trilete arm length to spore radius, trilete arms of unequal
length, presence and kind of exine surface ornamentation, number
and distribution of folds in exine, spore size range, and spore
shape. All of these features are assumed, by those who have
used them as a basis for Calamospora speciation, to be relatively
constant and not subject to significant developmental or diagenetic variation. It is the purpose of this report to examine
the range of variation shown by each of these characteristics in
Calamospora-like spores isolated from a number of calamitean
cone species, and to draw conclusions concerning the usefulness
of these features as taxonomic criteria.

MATERIALS AND METHODS

Cones yielding spores illustrated in this report were all
preserved in Pennsylvanian-age coal ball petrifactions. The
collection localities and cone species of illustrated specimens
are given below.

Cone species	Collection Localities
Calamostachys binneyana	Lewis Creek
Calamostachys inversibractis	Sahara
Litostrobus iowensis	Sahara
Pendulostachys cingulariformis	Berryville
Palaeostachya andrewsii	What Cheer, Sahara
Palaeostachya decacnema	Providence, Shade

Berryville locality - Sec. 7, T2N, R13W, Sumner 15' quad-
rangle, Lawrence County, Illinois. Stratigraphic position -
Calhoun Coal, Mattoon Formation, McLeansboro Group. Age- Late
Pennsylvanian.

Lewis Creek locality- Universal Transverse Mercator Grid
Coordinates $297,250E$ $4097,500N$, Cutchin 7.5' quadrangle, Leslie
County, Kentucky. Stratigraphic position- Copland Coal,
Breathitt Formation. Age- Early Pennsylvanian.

Providence locality- 87°46'14" West Longitude, 37°24'49"
North Latitude, Providence 7.5' quadrangle, Webster County,
Kentucky. Stratigraphic position- No. 11 or No. 12 Kentucky

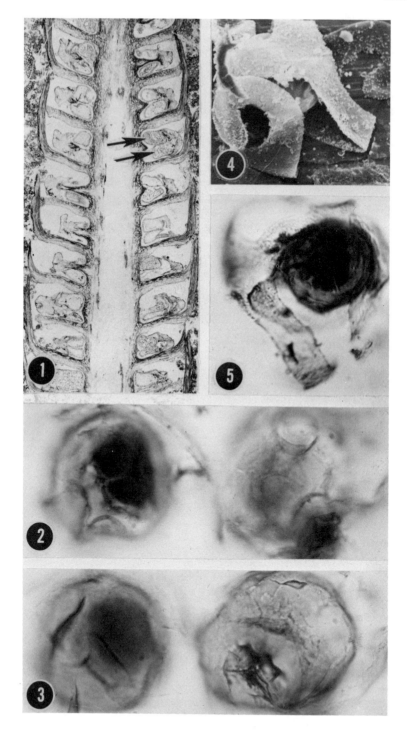

Figure 1. Median longitudinal section of a typical
calamitean cone (Calamostachys binneyana). Arrows indicate
spore-filled sporangia. Coal ball 421 Hl side, slide 102,
X 11.2.

Figure 2. Palaeostachya decacnema spores showing dark contact areas on lefthand spore. Compare with figure 3. Coal ball
224 Elc, maceration, X 1200.

Figure 3. Different focal plane of spores illustrated in
figure 2, showing spore on right without dark contact areas.
Coal ball 224 Elc, maceration, X 1200.

Figure 4. Scanning electron micrograph of Palaeostachya
decacnema spore showing attached elaters. Coal ball 6045, X 950.

Figure 5. Transmitted light micrograph of Elaterites-like
Palaeostachya decacnema spore. Coal ball 224 Elc, maceration,
X 950.

Coal, Carbondale Formation, Kewanee Group. Age- Middle
Pennsylvanian.

Sahara locality- Sec. 30, T9S, R4E, Harrisburg 15' quad-
rangle, Williamson County, Illinois. Stratigraphic position-
Herrin (No. 6) Coal, Carbondale Formation, Kewanee Group. Age-
Middle Pennsylvanian.

Shade locality- NW¼ NE¼ Sec. 21, T4N, R13W, Shade 7.5'
quadrangle, Athens County, Ohio. Stratigraphic position-
Pittsburgh (No. 8) or Redstone (No. 8a) Coal, Monogahela Series.
Age- Late Pennsylvanian.

What Cheer locality- NW¼ NE¼ Sec. 15, T76N, R13W, What Cheer
7.5' quadrangle, Kerokuk County, Iowa. Stratigraphic position-
Laddsdale Coal, (lower) Cherokee Group, Des Moines Series. Age-
Middle Pennsylvanian.

Details of the anatomy and morphology of each calamitean
cone species, as well as detailed descriptions of the spores
produced by each species have appeared in a previous publication
(Good, 1975). Litostrobus iowensis (not a calamitean cone) and
its spores have been described in a report by Baxter (1967).

Cellulose acetate peels were made of each specimen used in
this study in order to determine the internal anatomy and insure
correct cone species identification. Subsequently, spores were
macerated from intact sporangia using dilute (2%v/v) hydrochloric
acid. A small ridge of paraffin was constructed around each
macerated sporangium in order to minimize matrix contamination of
the spore sample.

POTENTIAL TAXONOMIC CHARACTERISTICS

Contact Areas. The presence or absence of darkened contact
areas (area contagionus) has been used as a major criterion in
delimiting certain Calamospora species. Among those species
described as lacking dark contact areas are C. pallida Schopf
et. al. (1944), C. flexilis and C. liquida Kosanke (1950), C.
obscura Peppers (1964), and C. decora Wilson and Hoffmeister
(1956). Species presumably characterized by darkened contact
areas include C. parva Guennel (1958), C. exigua Staplin (1960),
C. saarina and C. minuta Bhardwaj (1957), as well as C. hartungiana
Schopf et. al. (1944) and C. breviradiata Kosanke (1950).

The present investigation indicates that spores from a
single cone species and even a single cone specimen can show all
intermediates between very dark contact areas and contact areas
apparently lacking any differentiation at all. Figures 2 and 3

show different focal planes of the same two Palaeostachya
decacnema spores sitting side by side in the same maceration
sample. One spore lacks darkened contact areas while the other
spore has very dark regions between the trilete arms. The
circular area surrounding the trilete of the spore on the right
of figure 3 represents a partially uncoiled elater, not a partially differentiated contact area. Figures 14, 13, 16, and 17
show a series of Palaeostachya andrewsii microspores in which
darkening of the contact area is progressively less distinct.
In the present study, all cone species in which a large number
of specimens were available for examination showed variation in
the darkening of contact areas. Such variation has previously
been illustrated for spores of the calamitean cone genera
Macrostachya (Hartung, 1933), Palaeostachya, and Pendulostachys
(Good, 1975). The presence or absence of darkened contact areas
is due to spore ontogeny or diagenetic factors and does not appear
to be a valid taxonomic criterion. Thus, Calamospora species
such as C. hartungiana and C. liquida, which are said to be
similar to each other except for differences in contact area development (Kosanke 1950), may not in fact represent distinct
types of dispersed spores.

Globular Inner Body. Bhardwaj and Salujha (1963) define the
species C. exila as having a prominant globular inner body placed
eccentrically with respect to the trilete mark. Such globular
bodies have been observed in spores from several cone species
examined in the present study and possibly represent protoplasmic
contents. Illustrated spores showing such a globular inner body
include specimens from the cones Palaeostachya decacnema (Fig. 8)
and Pendulostachys cingulariformis (Fig. 15, left).

Separation of Exine Layers. At least one presumed species
of Calamospora (C. cavumis Shu, 1964) has been distinguished from
all other species by the presence of an exine that is separated
into presumed intexine and exoexine layers. Two spore genera,
Ricaspora Bhardwaj and Salujha (1963) and Perotriletes Couper
(1953) have been erected to include spherical trilete spores with
a perisporial covering. Ricaspora is reported from the Permian
(Bhardwaj and Salujha, 1963) and Pennsylvanian (Gupta 1970; Gupta
and Boozer, 1972). Although Perotriletes is mainly a Mesozoic
genus, it has been described from the Lower Carboniferous of
Europe (Hughes and Playford, 1961).

All calamitean microspores and isospores develop a perisporial covering which forms the elater layer (Figs. 2-5). It is
only when this perispore is lost that the spore takes on a
Calamospora-like configuration. It is thus probably not appropriate to distinguish separate species or genera of Calamospora-
like dispersed spores based upon the presence of an extra-exinous
layer.

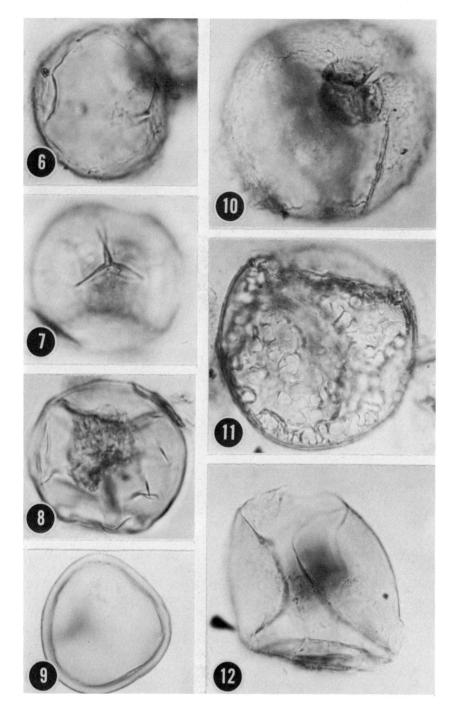

Figure 6. Calamospora-like spore from the sphenophyllalean cone Litostrobus iownsis. Coal ball 1039 K, maceration, X 1200.

Figure 7. Palaeostachya decacnema spore lacking dark contact areas. Compare with figures 2 and 5. Coal ball 6045 E, maceration, X 1200.

Figure 8. Palaeostachya decacnema spore with globular inner body possibly representing cytoplasmic contents. Coal ball 6045 H, maceration, X 1200.

Figure 9. Calamostachys inversibractis spore with thick exine. Compare with figure 12. Coal ball 2615 G, maceration, X 900.

Figure 10. Palaeostachya andrewsii microspore showing dark contact areas and Punctatisporites-like or Schopfites-like surface ornamentation. Coal ball 6010 H, maceration 1, X 765.

Figure 11. Schopfites-like ornamentation pattern on Palaeostachya andrewsii microspore. Coal ball 6010 H, maceration 1, X 765.

Figure 12. Flattened thin-walled Calamostachys inversibractis spore. Compare with thick exine of spore shown in figure 9. Coal ball 2615 H, maceration 2, X 850.

Wall Thickness. The genus Calamospora is generally defined
as having a very thin wall layer. Thick-walled spores that are
otherwise identical to Calamospora (smooth, spherical, trilete,
dark contact areas) are sometimes placed in the genus
Phyllothecotriletes Luber (1955), reported from the Devonian
(Luber, 1955), Mississippian (Staplin, 1960), and Pennsylvanian
(Barss, 1967) of Eurasia and North America. Some large (over
100 um.) dispersed Calamospora species are described as having
thick walls (C. obesus Schopf et. al., 1944; C. flava Kosanke,
1950; C. sinuosa Potonie and Kremp, 1955). However, there are
examples of large specimens with thin walls (C. perrugosus Schopf
et. al., 1944). Most species of Calamospora are defined as
having an exine that is "very thin" or that measures from 1-3 um.
in thickness. Except for the larger species mentioned above, no
variation in wall thickness outside of the contact areas is
given in any species diagnosis. Relative wall thickness has been
used in the literature to distinguish species (e.g. differentiation of C. membranata and C. saariana from C. pallida and C.
flexilis, and differentiation of C. minuta from C. brevirradiata
by Bhardwaj, 1957).

Exine thickness of spores from most calamitean cone species
examined is generally uniformly thin, measuring in many cases less
than one micrometer. However, spores of Calamostachys
inversibractis show a considerable variation in wall thickness.
In some spores the exine is too thin to accurately measure (Fig.
12), while other spores of this cone have exines up to 3 um. in
thickness (Fig. 9). Because such a large variation in wall thickness is known in at least one species of calamitean cone, the use
of wall thickness to distinguish various species of Calamospora
may not be appropriate. A reevaluation of criteria used to distinguish Phyllothecotriletes from Calamospora may show the two
spore types to be synonymous. Phyllothecotriletes is always
reported from rock units that contain specimens of Calamospora.

Ratio of Trilete Arm Length to Spore Radius. Most diagnoses
of Calamospora species note the length of trilete arms either in
absolute terms or as a ratio of arm length to spore radius. Such
ratios vary from 1/4 (C. parva Guennel, 1958) to 3/4 (C. obscura
Peppers, 1964). Recorded trilete lengths or ratios for most
Calamospora species are within rather narrow limits and have been
used (Bhardwaj, 1957; Staplin, 1960) to distinguish between
several presumably distinct species.

The present investigation indicates that trilete arm length
in spores from any particular cone species differs more than
descriptions of dispersed spore species would seem to suggest.
The ratio of arm length to spore radius can vary due to variations
in arm length and due to spore folding which decreases apparent

spore diameter. Figures 10, 13, 14, and 16 illustrate
Palaeostachya andrewsii microspores with increasingly longer
trilete arms and an increasingly smaller arm length to spore
radius ratio (Fig. 10, ratio 1/3; Fig. 16, ratio 4/5). Calamitean
spores from the cone Calamocarpon insignis have previously been
illustrated with trilete rays varying in length by 50% (Good and
Taylor, 1974). Dispersed spores identified as the same species
have been illustrated by Guennel (C. pallida, 1958) and Samoilovich
(C. hartungiana, 1953) with great variation in trilete arm length.
Trilete arm length does not appear to offer an adequate criterion
for distinguishing dispersed species within Calamospora.

     Unequal Length of Trilete Arms. Calamospora breviradiata
Kosanke (1950), C. straminea Wilson and Kosanke (1944; as described by Smith and Butterworth, 1967), and C. exigua Staplin
(1960) are all described as frequently having trilete arms of
unequal length.

     If enough spores are examined, some spore specimens from
all calamitean cone species studied are found with unequal
trilete arms. Figures 13, 16, and 18 illustrate Palaeostachya
andrewsii microspores with unequal trilete arms, while figures
10 and 14 show microspores of this same cone species in which
each arm of a trilete is approximately the same length.

     Presence and Kind of Surface Ornamentation. Spores of the
Calamospora type are generally believed to lack any appreciable
surface ornamentation. Punctate spherical trilete spores lacking
dark contact areas are placed in the dispersed genus
Punctatisporites Schopf et. al. (1944). Kosanke (1950) notes
that mildly ornamented species of this genus might be confused
with Calamospora, and indeed there has been some switching of
dispersed species back and forth between the two genera.
Calamospora exigua (Staplin, 1960), C. hartungiana Schopf et. al.
(1944), and C. decora Wilson and Hoffmeister (1956) are currently
recognized as having minutely to coarsely granular or punctate
walls.

     Wall ornamentation in spores of all cone species examined
varies from smooth to minutely punctate (compare Fig. 9 with
Fig. 12; Fig. 13 with Figs. 14, 16, and 18). Such minute ornamentation does not appear to adequately distinguish between
spores produced by different species of calamitean cones.

     In two cone species, Calamostachys binneyana and
Palaeostachya andrewsii, certain cone specimens yield spores with
various combinations of smooth and grossly ornamented areas
(Figs. 10, 11). Sculptured areas have a "moth eaten" appearance.
Ornamentation appears irregularly reticulate and often includes

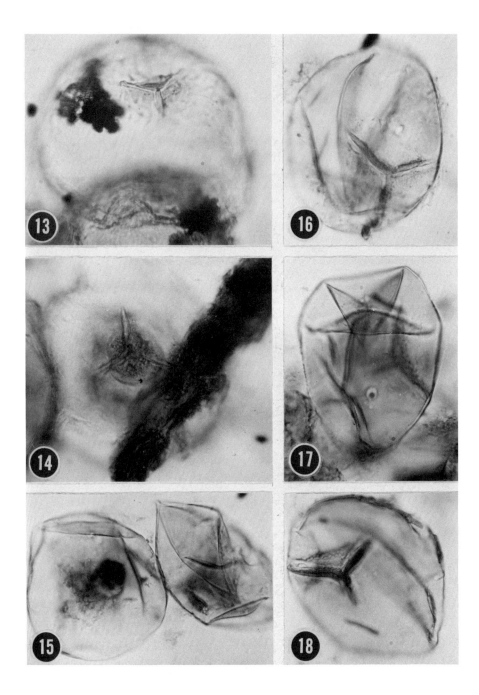

Figure 13. <u>Palaeostachya</u> <u>andrewsii</u> microspore without dark contact areas and with trilete rays of uneven length. Coal ball 5466 A, maceration, X 765.

Figure 14. <u>Palaeostachya</u> <u>andrewsii</u> microspore with dark contact areas and with trilete rays of equal length. Coal ball 6010 H, maceration 1, X 765.

Figure 15. <u>Pendulostachys</u> <u>cingulariformis</u> spores showing globular inner body (left), folds parallel to spore margin (left), and criss-cross folds (right). Coal ball 4122 H, maceration, X 765.

Figure 16. <u>Palaeostachya</u> <u>andrewsii</u> microspore with folds parallel to spore margin. Coal ball 837 F, maceration, X 765.

Figure 17. <u>Palaeostachya</u> <u>andrewsii</u> microspore with criss-cross fold pattern. Coal ball 837 F, maceration, X 765.

Figure 18. <u>Palaeostachya</u> <u>andrewsii</u> microspore with weakly developed contact areas, trilete rays of unequal length, and a single major longitudinal fold. Coal ball 837 F, maceration, X 765.

apparent perforations of the exine and large flattened upward
projections (Fig. 11). Such highly ornamented calamitean spores
appear to correspond most closely with the dispersed spore genus
Schopfites Kosanke (1950). This genus delimits radial trilete
spores with both smooth and highly ornamented areas of the exine.
Four Schopfites species have been described by Kosanke (1950),
Bhardwaj (1957), and Habib (1960). Individual Calamostachys
binneyana and Palaeostachya andrewsii spores have been noted in
the present investigation which are similar to each of the four
species. In the literature, Schopfites is always reported in
association with Calamospora.

Ornamentation on Schopfites-like calamitean spores is
probably diagenetically induced, since typical spores of C.
binneyana and P. andrewsii have smooth exines. Neves and Sullivan
(1964) have shown how pyrite crystals can cause perforations in
the exine of fossil pollen and spores. These authors illustrate
a perforate Calamospora specimen somewhat resembling that shown
here in Figure 11. Moore (1963) has described smooth spores with
an apparent branching network caused by the presence of a
saprophytic organism which had apparently begun to destroy the
exine prior to fossilization.

The presence of Schopfites in a spore flora has been considered by some authorities to be stratigraphically significant,
with Schopfites considered a marker or index genus for certain
horizons (Barss et. al., 1963; Gray and Taylor, 1967; Gupta,
1970; Kosanke 1950). Results of the present investigation suggest
that a reevaluation of Schopfites and its importance as a stratigraphic indicator would be appropriate.

Folds in Exine. Because of the thin wall of most calamitean
spores, many exhibit various types of folds in the exine. These
folds have been called "compression" folds in the palynological
literature. However, the fact that many spores from calamitean
cones preserved in coal balls are highly folded suggests that
many folds are not due to compression of overlying sediments, but
due instead to spore abortion or dehydration prior to fossilization.

Several described Calamospora species are said to be characterized by various types of folding. Habib (1966) describes C.
multiplicata as characterized by criss-cross folds and easily
distinguished from the otherwise similar C. breviradiata which
has folds parallel to the spore margin. Peppers (1970) was unable
to distinguish between these two species on the basis of folding.
Kosanke (1950) describes species characterized by folds associated
with the trilete (C. flexilis), by numerous folds parallel to the
spore outline (C. liquida), and by a single major fold (C. pedata).
Calamospora plicata Tiwari and Navale (1967) bears many parallel

folds. The type species of the genus, C. hartungiana is characterized by "several lenticular folds" (Schopf et. al., 1944).

There is no general pattern for spore folding observed in
collapsed spores isolated from sporangia of the cone species
examined for this study. The size, number, and distribution of
folds in spores from each cone species is quite variable, and
all imaginable fold configurations are present. Figure 15
illustrates parallel and criss-cross folding in two spores of
Pendulostachys cingulariformis. Microspores of Palaeostachya
andrewsii may be spherical and lack folds (Figs. 10, 13, 14),
have a single longitudinal fold (Fig. 18), have a series of folds
parallel to the spore surface (Fig. 16), or have folds which
criss-cross each other (Fig. 17). The fact that such a wide
variety of folding patterns are found in spores produced by the
same cone suggests that fold configurations should not be a
basis for speciation in dispersed Calamospora-like spores.

Spore Size Range. Within a period of three years three new
Calamospora species were each described as averaging 37-38 um.
in diameter and as being the smallest known species of the genus
(C. diversiformis Balme and Hennelly, 1956; C. minuta Bhardwaj,
1957; C. parva Guennel, 1958). Subsequently, an additional
species has been described on the basis of small size (C. pusilla
Peppers, 1964, 30 um. average diameter). According to Winslow
(1959), large Calamospora-like spores, presumably megaspores, are
generally placed in C. laevigata Schopf et. al. (1944) or C.
sinuosa Potonie and Kremb (1955) depending almost entirely upon
whether the spore is large (250-450 um. for C. laevigata) or very
large (600-1050 um. for C. sinuosa). Descriptions of other
Calamospora species all include spore size, with the reported
maximum to minimum size range extending through approximately
10-20 um. in most species.

Size ranges and size frequency distributions for the isospores,
microspores, and megaspores of cones studied in the present investigation have been published elsewhere (Good, 1975). Spore
size range for cone species is greater than that given for most
dispersed species of Calamospora, and the size ranges of spores
produced by different cones overlap each other. Size along does
not adequately differentiate between spores produced by most
cone species and probably should not be used as a basis for the
establishment of dispersed Calamospora species.

Spore Shape. Calamospora represents spores which were
originally spherical, and most species are described as having
a circular amb. However, exine folding can alter the original
circular outline of the spore, and a number of Calamospora
species have been described as characterized by folding that

results in an outline that is other than circular. Calamospora
flexilis Kosanke (1950) is said to be roundly triangular. Both
C. multiplicata and C. elliptica Habib (1966) have a characteristic eliptical outline. Calamospora pseudotriangulara Habib
(1966) has a triangular outline due to large folds. The type
species, C. hartungiana Schopf et. al. (1944) is described as
generally compressed to a polygonal or lenticular outline.

The present investigation suggests that, just as spore
folding shows infinate variability in a single cone species, the
resulting outline of folded spores from a single cone species
is also highly variable. Figure 15 shows Pendulostachys
cingulariformis spores that are both circular and lenticular.
Palaeostachya andrewsii microspores are illustrated which are
spherical (Figs. 10, 13, 14), roundly elliptical (Fig. 16),
lenticular-elliptical (Fig. 18), and polygonal-triangular (Fig.
17).

GENERAL DISCUSSION

Although Calamospora is often considered as having a
natural affinity to the plant Calamites, not all Calamospora-like
dispersed spores are referable to this plant. The presence in a
spore flora of dispersed Calamospora specimens lacking darkened
contact areas is not, by itself, sufficient evidence for the
presence of Calamites. Cones of the Noeggerathiales, including
Discinites, (Arnold, 1944, 1949) are known to produce this spore
type. In addition, cones from some members of the Sphenophyllales
can produce spores which would appear Calamospora-like if found
dispersed. The spiny and the operculate (Vestispora-like) spores
of some Sphenophyllum cones have an inner spore body that would
appear Calamospora-like in the absence of the spiny or operculate
perispore. Such a Calamospora-like spore is illustrated in figure
6 from the sphenophyllalean cone Litostrobus iowensis.
Sphenophyllum cones are not known to produce spores with dark
contact areas.

Variation found in spores from the same calamitean cone
species and within the same calamitean cone sporangium suggests
that the criteria discussed above, whether used singly or in
combination, cannot distinguish between elaterless Calamospora-
like spores produced by different cone species. Wilson (1962) has
noted that dispersed Calamospora species are only form species
defined on observable easily recognizable taxonomic characters
and do not necessarily represent fossil spores of a particular
natural plant species. The genus Calamospora is fairly easy to
recognize in the dispersed state. However, the present study
clearly suggests that morphologic features of calamitean spores
do not easily allow a clearcut division of species within this

dispersed spore genus. The confusing taxonomy within Calamospora
has been noted in several palynological investigations (Dijkstra,
1957; Felix and Burbridge, 1967; Smith and Butterworth, 1967).
When a large number of dispersed Calamospora spores are studied,
these authors indicate that the definite segregation of spore
species is nearly impossible.

Stratigraphic interpretations based upon different species
of Calamospora may be invalid. Fortunately, the literature contains very few examples of such interpretations. Winslow (1959)
suggests that the presumed megaspore C. sinuosa has a restricted
stratigraphic distribution. Some palynological stratigraphic
studies have drawn conclusions based in part upon the presence
and frequency of the genus Calamospora without making reference
to individual species. Hacquebard and Barss (1957) suggest that
the low frequency or absence of Calamospora is characteristic of
Mississippian spore floras. Upshaw and Creath (1965) have used
the presence of this genus as partial evidence for the
Pennsylvanian age of a presumed cave deposit found in Devonian
limestone. Additional studies in which the genus Calamospora
is considered to have some stratigraphic importance are those of
Bhardwaj (1955), Coquel et. al. (1970), and Habib (1968). Although Calamospora has a very wide stratigraphic range (Devonian -
Triassic), these authors feel that the relative abundance of this
genus in a dispersed spore flora can prove useful in stratigraphic
palynology.

In previous reports (Good, 1975; Good and Taylor, 1975) it
has been noted that distinctions between the dispersed spore
genera Vestispora, Elaterites, Ricaspora, Perotriletes, and
Calamospora may be more apparent than real, since calamitean
spores go through developmental stages resembling all these
genera. The present report adds Phyllothecotriletes and Schopfites
to the list of dispersed genera resembling spores produced by
calamitean cones. Stratigraphic interpretations based upon
apparent floral differences among any of these genera should
probably be made with caution. It is significant that carboniferous-
age Vestispora, Elaterites, Ricaspora, Perotriletes,
Phyllothecotriletes and Schopfites specimens are almost always
described as minor components of spore floras which contain a
much higher percentage of Calamospora-like sporomorphs.

LITERATURE CITED

Abbott, M. L. 1968. Lycopsid stems and roots and Sphenopsid fructifications and stems from the Upper Freeport Coal of southeastern Ohio. Palaeontographica Amer., 6(38): 1-49.

Arnold, C. A. 1944. A heterosporous species of Bowmanites from the Michigan coal basin. Amer. J. Bot., 31: 466-469.

Arnold, C. A. 1949. Fossil Flora of the Michigan coal basin. Contr. Mus. Paleont. Univ. Michigan, 7(9): 131-269.

Balme, B. E. and J. P. F. Hennelly. 1956. Trilete sporomorphs from Australian Permian sediments. Australian J. Bot., 4: 240-260.

Barss, M. S. 1967. Illustrations of Canadian fossils. Carboniferous and Permian spores from Canada. Geol. Surv. Canada Paper 67-11.

Barss, M. S., P. A. Hacquebard, and R. D. Howie. 1963. Palynology and stratigraphy of some Upper Pennsylvanian and Permian rocks of the Maritime Provinces. Geol. Surv. Canada Paper 63-3.

Baxter, R. W. 1967. A revision of the sphenopsid organ genus Litostrobus. Univ. Kansas Sci. Bull. 48: 1-23.

Bhardwaj, D. C. 1955. The spore genera from the Upper Carboniferous coals of the Saar and their value in stratigraphic studies. Palaeobotanist 4: 119-149.

Bhardwaj, D. C. 1957. The palynological investigations of the Saar coals (Part 1. Morphology of Sporae dispersae). Palaeontographica B, 101: 73-125.

Bhardwaj, D. C. and S. K. Salujha. 1963. Sporological study of seam VIII in Raniganj coal field, Bihar (India) - Part 1. Description of sporae dispersae. Palaeobotanist 12: 181-215.

Coquel, R., S. Loboziak, and Y. Lemoigne. 1970. Confirmation de l'âge Westphalien du houiller de Le Plessis (Manche) d'après l'étude palynologique de quelques échantillons de charbon. Ann. Soc. Geol. Nord, 90: 15-21.

Couper, R. A. 1953. Upper Mesozoic and Cainozoic spores and pollen grains from New Zealand. N. Zealand Geol. Surv. Paleont. Bull. 22.

Delevoryas, T.   1955.   A Palaeostachya from the Pennsylvanian of
    Kansas.   Amer. J. Bot. 42: 481-488.

Dijkstra, S. J.   1957.   Lower Carboniferous megaspores.
    Geologische Sticht. Mededeelingen. New Series C III.
    1 Palaeobotanie 10: 5-18.

Felix, C. J. and P. P. Burbridge.   1967.   Palynology of the Springer
    Formation of southern Oklahoma, U.S.A. Palaeontology, 10:
    349-425.

Good, C. W.   1975.   Pennsylvanian-age calamitean cones, elater-
    bearing spores, and associated vegetative organs.
    Palaeontographica B, 153: 28-99.

Good, C. W. and T. N. Taylor.   1974.   The establishment of
    Elaterites triferens spores in Calamocarpon insignis
    microsporangia. Trans. Amer. Micros. Soc. 93: 148-151.

Good, C. W. and T. N. Taylor.   1975.   The morphology and systematic
    position of Calamitean elater-bearing spores. Geoscience
    and Man, 11: 133-139.

Gray, L. R., and T. N. Taylor.   1967.   Palynology of the
    Schultztown Coal in western Kentucky. Trans. Amer.
    Microsc. Soc. 86: 502-506.

Guennel, G. K.   1958.   Miospore analysis of the Pottsville coals
    of Indiana. Indiana Dept. Conservation Geol. Surv. Bull. 13.

Gupta, S.   1970.   Miospores from the Desmoinesian-Missourian
    boundary   Formations of Texas and the age of the Salesville
    Formation. Geoscience and Man, 1: 67-82.

Gupta, S., and O. W. Boozer.   1972.   Palynology of the Garner
    Formation (Strawn Series) of north-central Texas.
    Geoscience and Man 4: 119-125.

Habib, D.   1966.   Distribution of spore and pollen assemblages in
    the Lower Kittanning Coal of western Pennsylvania.
    Palaeontology 9: 629-666.

Habib, D.   1968.   Spore and pollen paleoecology of the Redstone
    seam (Upper Pennsylvanian) of West Virginia.
    Micropaleontology, 14: 199-220.

Hacquebard, P. A., and M. S. Barss.   1957.   A Carboniferous spore
    assemblage in coal from the South Nahanni River area,
    Northwest Territories. Geol. Surv. Canada Bull. 40.

Hartung, W. 1933. Die Sporenverhältnisse der Calamariaceen.
    Prussia. Geologische Landesanstalt. Institut fur
    Palaobotanik und Petrographie der Brennsteins. Arbeiten.
    3: 96-149.

Hoffmeister, W. S., F. L. Staplin, and R. E. Malloy. 1955.
    Geologic range of Paleozoic plant spores in North America.
    Micropaleontology 1: 9-27.

Hughes, N. F. and G. Playford, 1961. Palynological reconnaissance
    of the Lower Carboniferous of Spitsbergen. Micropaleontology
    7: 27-44.

Kosanke, R. M. 1950. Pennsylvanian spores of Illinois and their
    use in correlation. Ill. St. Geol. Surv. Bull. 74.

Luber, A. A. 1955. Atlas of spores and pollen grains of the
    Paleozoic deposits of Kazakhstann (translation). Akad.
    Sci. Kazakh., S.S.S.R., Alma Ata.

Mädler, K. 1964. Die geologische Verbreitung von Sporen und
    Pollen in der Deutschen Trias. Beih. geol. Jb. 65: 7-147.

Moore, L. R. 1946. On the spores of some Carboniferous plants;
    their development. Geol. Soc. London Quart. J. 102: 251-
    298.

Moore, L. R. 1963. Microbiological colonization and attack on
    some Carboniferous microspores. Palaeontology 6: 349-372.

Neves, R. and H. J. Sullivan. 1964. Modification of fossil
    spore exines associated with the presence of pyrite crystals.
    Micropaleontology 10: 443-452.

Peppers, R. A. 1964. Spores in strata of Late Pennsylvanian
    cyclothems in the Illinois Basin. Ill. St. Geol. Surv.
    Bull. 90.

Peppers, R. A. 1970. Correlation and palynology of coals in
    the Carbondale and Spoon Formations (Pennsylvanian) of the
    northeastern part of the Illinois Basin. Ill. St. Geol.
    Surv. Bull. 93.

Potonie, R., and G. Kremp. 1955. Die sporae dispersae des
    Ruhrkarbons, ihre Morphographie und Stratigraphie mit
    Ausblicken auf Arten anderer Gebiete und Zeitabschnitte:
    Teil I. Palaeontographica B, 98: 1-136.

Samoilovich, S. R. 1953. Pollen and spores from the Permian deposits of the Cherdyn' and Akyubinsk areas, Cis-Urals. Paleobotanicheskii sbornik: Vsesoiuznyi nauchno-issledovatel'skii geologo-ravedochnyi institut, Leningrad, Trudy, new series, 75: 5-57. (English translation, 1961, Oklahoma Geol. Surv. Circ. 56).

Schopf, J. M., L. R. Wilson, and R. Bentall. 1944. An annotated synopsis of Paleozoic Fossil Spores and the definition of generic groups. Ill. St. Geol. Surv. Rept. Invest. 91.

Shu, O. 1964. A preliminary report on sporae dispersae from the Lower Shihhotze Series of Hoku District, NW Shansi. ACTA Palaeont. Sinica, 12: 486-519.

Smith, A. H. V., and M. A. Butterworth. 1967. Miospores in the coal seams of the Carboniferous of Great Britain. Special papers in palaeontology No. 1. Palaeontological Association, London.

Staplin, F. L. 1960. Upper Mississippian plant spores from the Golata Formation, Alberta, Canada. Palaeontographica B, 107: 1-40.

Taylor, T. N. 1967. On the structure of Calamostachys binneyana from the lower Pennsylvanian of North America. Amer. J. Bot., 54: 298-305.

Tiwari, R. S. and G. K. B. Navale. 1967. Pollen and spore assemblage in some coals of Brazil. Pollen et Spores, 9: 583-605.

Upshaw, C. F. and W. B. Creath. 1965. Pennsylvanian miospores from a cave deposit in Devonian limestone, Callaway County, Missouri. Micropaleontology, 11: 431-448.

Wilson, L. R. 1943. Elater-bearing spores from the Pennsylvanian strata of Iowa. Amer. Mid. Nat., 30: 518-523.

Wilson, L. R. 1962. Permian plant microfossils from the Flowerpot Formation, Greer County, Oklahoma. Oklahoma Geol. Surv. Bull. 49.

Wilson, L. R., and W. S. Hoffmeister. 1956. Plant microfossils of the Croweburg Coal. Oklahoma Geol. Surv. Circ. 32.

Wilson, L. R. and R. M. Kosanke. 1944. Seven new species of unassigned plant microfossils from the Des Moines Series of Iowa. Proc. Iowa Acad. Sci., 51: 329-333.

Winslow, M. R.   1959.   Upper Mississippian and Pennsylvanian
    megaspores and other plant microfossils from Illinois.
    Ill. St. Geol. Surv. Bull. 86.

    Slides bearing all illustrated specimens are deposited with
the paleobotanical collection, Department of Botany, Ohio State
University, Columbus, Ohio, 43210.

DEPOSITIONAL AND FLORISTIC INTERPRETATIONS OF A POLLEN DIAGRAM FROM

MIDDLE EOCENE, CLAIBORNE FORMATION, UPPER MISSISSIPPI EMBAYMENT

Frank W. Potter, Jr.

Department of Plant Sciences, Indiana University

Bloomington, Indiana U.S.A. 47401

ABSTRACT

A middle Eocene pollen profile was extracted from a 9-meter series of clay and lignite in western Tennessee. Recent pollen depositional information on small basins was used to interpret depositional environments and pollen source areas. Two depositional systems were operant during deposition, an open system and a closed system. Using the change in depositional systems, three pollen source areas, local, background, and regional were established. The results show care must be taken in stratigraphic correlation of small nonmarine fossil deposits when this is based on types present and relative abundance. Plant communities of the three source areas varied significantly in composition, a situation which must be considered in floristic interpretation of past community structures and paleoclimates.

INTRODUCTION

Recent examination of modern pollen rain, its subsequent deposition into small basins (Tauber 1967, Berglund 1973, Flenley 1973, Janssen 1966, 1973) and the relative importance of waterborne over windborne grain transport, provide a new basis on which to evaluate fossil pollen assemblages from small, nonmarine deposits. The following investigation was conducted to determine what, if any, vertical fluctuation occurs within small plant fossil deposits in the upper portion of the Mississippi Embayment. It is necessary to establish whether vertical change occurs at the sites before accurate stratigraphic correlation among the many deposits of the area can be attempted. Palynological evidence contained within the

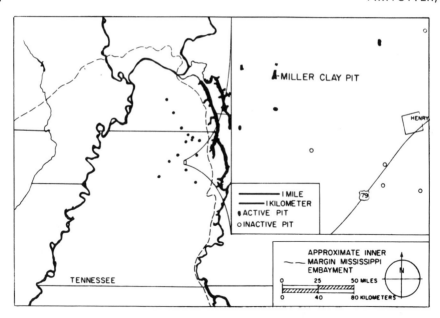

Fig. 1. Fossil localities in western Kentucky and Tennessee.

vertical fluctuation would provide information on past plant
community structure and composition.

Within the Mississippi Embayment region of southeastern North
America, numerous plant fossil rich clay and lignite deposits are
located. A group of small, disjunct, nonmarine lenses of clay with
associated lignites occur in the upper portion of the Embayment
(Fig. 1). The fossil sites are situated along the eastern margin
of the Embayment in western Kentucky and Tennessee. The majority,
if not all, are assignable to the Middle Eocene, Claiborne Formation (Dilcher 1971, Parks 1971, Potter 1976).

Plant fossil remains were first noted by Lesquereux (1859) and
later investigated extensively by Berry (1914, 1916, 1924, 1930,
1937, 1941). Major reexamination of the floras has been undertaken
by Bilcher and other workers. Dilcher (1971, 1973) emphasizes the
excellent preservation of the Eocene plant fossils, considering the
locations some of the best Early Tertiary fossil flora sites in the
world. Only a few palynological investigations have been carried
out on the Middle Eocene sediments of the upper Mississippi Embayment area (Tschudy 1973, Elsik 1974). These investigations have
attempted a correlation of deposits scattered over the entire Embayment, based on occurrence of a few types and their relative abundance. Sufficient data is not available for corroboration of a

technique that employs palynomorph types on such a large scale and involves different regional as well as local pollen sources. Unfortunately, time generally permits only one or a few samples to be examined per locality, effectively eliminating the possibility of within-site variation from consideration.

MILLER CLAY PIT

Miller Clay Pit, located 3.6 miles west-northwest of Henry, Tennessee (Henry Quadrangle, 36°13'15" N. Lat., 88°26'46" W. Long.) was sampled to investigate possible vertical pollen fluctuation. The pit has been actively mined for several years by the H. C. Spinks Clay Company which has provided information on size and shape of the deposit. The site apparently represents the fillings of an abandoned channel feature on an ancient meandering stream floodplain (Dilcher 1971, Potter 1976).

The lithology of the pit indicates several changes in the depositional environment. Between 60 cm. to 1 meter of tan clay void of fossils forms the base of the deposit. The base clay abruptly changes into a dark brown to black clay which reaches a maximum thickness of 470 cm. Abundant plant mega- and microfossils are present in this dark clay zone. The zone grades rapidly into 290 cm. of lignite. On top of the lignite zone are patches of tan clay, from 80 cm. to 1 meter in depth and sterile of fossils. The dark clay and lignite provide no evidence of the laminations characteristic of seasonal flooding. The leaf compressions are denser toward the upper portion of the dark clay zone, reflecting a decrease in sedimentary rates.

PALYNOLOGICAL TECHNIQUES

A vertical series of samples was collected from the different lithologic units at Miller Clay Pit. These samples were taken at 10 cm. intervals throughout 80 cm. of basal tan clay, 470 cm. of dark clay, 290 cm. of lignite, and 80 cm. of the overlying tan clay. In the laboratory 2 cm^3 of each sample was chemically processed. The process utilized standard procedures of HF treatment to remove clay particles, oxidation with Schulze reagent, and 5% KOH, and acetolysis (for detailed field and laboratory procedures, see Potter 1976).

After slide preparation, grains were identified and 500 grains counted per level. Pollen percentages were calculated based on 71 grain types.

RESULTS

A diverse microfossil assemblage was recovered from the dark brown clay and lignite zones at Miller Clay Pit. Seventy-one pollen grain types were selected for analysis of distribution pattern. Additional types occurred; however, these were extremely rare and consequently showed no discernible distribution pattern in the profile. Spores were not included in the analysis of the diagrams; fungal spores varied with amount of organic material present in the sample, and other spores were rare and scattered throughout the vertical sequence.

Based on the lithologic change of sediment, two zones were distinguishable, the dark brown clay and the overlying lignite. Similarly, the two zones are separable by palynomorph content, although they represent a continuous depositional unit. The profile of the 71 selected pollen types shows the clay zone to be approximately twice as diverse as the lignite (Table 1). A maximum of 51 types of the 71 grains were recorded at one clay level, dropping to a minimum of 32 types. In comparison, the lignite ranged from 21 to 41 types. If only those pollen types representing over one percent of the total count per level is considered, the lignite reached a low of three grains at one level, with the clay containing the maximum of 21 at one level.

Examination of the distribution pattern of the pollen (Table 2) indicates that the clay is characterized by 42 types, the lignite by 12 types. Seventeen pollen types showed no tendency to either zone, their distribution and relative abundance independent of lithologic control. Two genera dominated both zones, Cupuliferoidaepollenites spp. and Cupuliferoipollenites spp. Cupuliferoidaepollenites spp. are dominant in the clay zone, reaching a maximum relative abundance of over 60 percent, dropping to minor significance in the lignite. Cupuliferoipollenites spp. show the opposite distribution, the grains reaching a maximum representation of just over 85 percent in the lignite. All of the other pollen types are represented by 10 percent or less of the total count per level in both zones

TABLE 1.   Diversity of pollen grains.

ZONE	Total diversity per level		Diversity of grains over 1% per level	
	Range	Mean	Range	Mean
Lignite	21 to 41	28	3 to 11	7
Clay	32 to 51	43	9 to 21	14

TABLE 2.   Pollen types and distribution patterns.

Pollen Type	Clay Zone	Lignite Zone
Cupuliferoidaepollenites sp. (14 to 19μ size)	***	**
Cupuliferoidaepollenites sp. (20 to 24μ size)	*	--
Momipites (coryloid form, 14 to 18μ size)	*	--
Momipites (coryloid form, 19 to 24μ size)	*	--
Tricolpites sp. (microreticulate form)	*	--
Tricolpopollenites sp. (granular form)	*	--
Tricolpopollenites retiformis	*	--
Nyssapollenites sp.	--	
Diporate sp. A	*	--
Tricolpate sp. A	--	
Liliacidites sp. (spherical form)	--	
Tricolpites geranoides	--	
Tricolpites sp. (reticulate form E)	--	
Tricolpopollenites sp. (verrucate form)	--	
Tricolpopollenites sp. (microspherical form)	--	
Subtriporopollenites sp.	--	
Multiporopollenites sp.	--	
Triporopollenites sp. (corylus form)	--	
Tricolpites sp. (reticulare form F)	--	
Tricolpopollenites hians (Form 1)	--	
Tricolpites sp. (reticulate form D)	--	
Tricolpites sp. (reticulate form C)	--	

Pollen Type	Clay Zone	Lignite Zone
Tricolpites sp. (reticulate form A)	--	
Tricolpopollenites hians (Form 2)	--	
Tricolpopollenites sp. (rugulate form)	--	
Tricolpate sp. B	--	
Tricolpites sp. (reticulate form B)	--	
Betulaceoipollenites sp.	--	
Ulmipollenites sp.	--	
Multiaperturate sp. A	--	
Triporopollenites sp. (equatorial pore form)	--	
Multiaperturate sp. B	--	
Triporopollenites pliktosus	--	
Tricolpites brevicolpus	--	
Tricolpites matauraensis	--	
Quercoidites sp.	*	
Tricolpate sp. C	*	--
Foveotricolpites prolatus	*	--
Tricolpites fissilis	--	
Aesculiidites variabilis (colpate form)	--	
Tricolporate sp. A	--	--
Pollenites ventosus	--	--
Tricolpate sp. D	--	--
Momipites sp. (coryphaeus form)	--	
Sapotaceoidaepollenites sp. (20 to 25μ size)	--	--

Pollen Type	Clay Zone	Lignite Zone
Sapotaceoidaepollenites sp. 　　(35 to 40μ size)	--	--
Tricolporopollenites sp. (form G)	--	--
Multiaperturate sp. C	--	--
Tricolporopollenites sp. (form A)	--	--
Cupuliferoidaepollenites sp. 　　(23 to 33μ size)	--	--
Tricolporopollenites sp. (form B)	*	*
Tricolporopollenites sp. (form C)	*	*
Tricolporopollenites sp. (form D)	*	*
Tricolporopollenites sp. (form E)	--	--
Aesculiidites variabilis (colporate form)	--	--
Tricolporate sp. B	--	--
Myrtaceidites parvus	--	--
Tricolporate sp. D	--	--
Tricolporopollenites sp. (form H)	--	--
Cupuliferoipollenites cingulum (Form 1)	**	***
Cupuliferoipollenites cingulum (Form 2)	--	*
Tricolporate sp. C	--	*
Cupuliferoipollenites sp.	--	*
Tricolporopollenites sp. (form F)		--
Extratriporopollenites terminilis		--
Liliacidites variegatus		*
Liliacidites dividuus		*
Monocolpopollenites tranquillus		*

Pollen Type	Clay Zone	Lignite Zone
Arecipites sp.		--
Sabalpollenites sp.		--
Anacolosidites efflatus		--

```
*   =  1 to 10% relative abundance
**  =  10 to 20% relative abundance
*** =  Over 20% relative abundance
--  =  1% or less and/or constantly present
```

DISCUSSION AND CONCLUSION

Three major pollen distribution patterns are discernible from the vertical sequence of Miller Clay Pit. These distributions are represented by (1) clay zone controlled grains; (2) lignite zone controlled grains; and (3) grains distributed independently of lithologic zones. The depositional system forming the clay and lignite is important in interpreting these patterns.

The clay zone, an open depositional system, represents sedimentation of locally derived organic and inorganic material and transported materials by overbank flooding. In such a depositional environment a high ratio of inorganic particles to organic matter would result, i.e., clay deposits. Pollen present would represent local source vegetation as well as both waterborne and windborne grains. The lignite zone represents a closed depositional system. The majority of the sediment materials are locally derived, predominantly organic matter. Pollen deposition in the closed system would originate from local vegetation source areas and windborne distant grains. The clay and lignite sediments at Miller Clay Pit are believed to represent 500 to 1000 years of deposition. This time span is based on sedimentary rates for similar modern, temperate environments. Rates of deposition are unfortunately not available for corresponding small basin depositional environments in tropical or subtropical conditions; however, it is doubtful that the time span exceeded the estimates given here.

The type of depositional system operating controls the method of pollen transport into the basin. Crowder and Cuddy (1973) assess the change from waterborne to windborne as a major factor in assemblage change in a modern, temperate system.

The pollen distribution pattern, rate of deposition, and method of transport indicate the existence of three source areas; these are (1) local vegetation adjacent to the depositional basin; (2) background vegetation, i.e., plants that are part of the surrounding floodplain environment although they do not immediately surround the basin, and (3) regional vegetation.

The lignite zone controlled grains generally represent the local vegetation source area. The diversity of types is relatively low compared to the total assemblage (12 of the 71 quantitatively-analyzed types), and the grains present are dominated almost exclusively by one grain, Cupuliferoipollenites spp. Six pollen types in the lignite show an increase in the upper portion of the zone, one of which, Anacolosidites sp., was exclusively confined to this portion. The other five grains are monosulcates represented by Liliacidites variegatus, L. dividuus, Monocolpopollenites tranquillus, Arecipites sp., and Sabalpollenites sp.; several have affinity with Palmae. These six pollen types apparently indicate a successional stage of the lowland vegetation surrounding the basin. Similar successional stages occur in modern lowland areas in tropical to subtropical/warm temperate climates. This stage is probably controlled by edaphic conditions and as such should not be used as a climate indicator.

The background vegetation is represented by the lithologically independent grains. These grains are derived from a more successionally stable plant community on the floodplain. While rare, they are consistently present throughout the vertical series. Investigations of small basin pollen deposition in temperate settings (Tauber 1967, Berglund 1973) and tropical situations (Flenley 1973) account for the movement of background vegetation pollen into the basins as primarily windborne from a relatively short distance, not greater than a few kilometers.

Regionally derived pollen generally is represented by the numerous rare grains found in the clay zone; in the lignite their occurrence is only scattered as single counts, or they are completely absent. The grains were transported predominantly by water entering the basin during the open depositional system phase, with occasional representation of windborne grains accounting for the scattered occurrence in the lignite. Muller (1959) records similarly distributed grains in deltaic levee deposits on the shelf of the Orinoco River system. Alnus, an extreme example, was waterborne more than 500 miles from its source. Several pollen types represented by the fossil pollen Abietineaepollenites, Podocarpidites, and Ephedra were recorded rarely and in a scatter distribution throughout the complete section and are regarded as windborne examples of regionally derived grains. Such types were not included in the quantitative analysis.

Other clay zone grains, which show a two- to tenfold decrease in the lignite, represent floodplain vegetation located along incoming waterways. During the clay deposition phase the pollen was transported by water and wind. With the elimination of the waterborne pollen as the lignite accumulated, the sediments increasingly reflected the background vegetation.

Three vegetation source areas are discernible for Miller Clay Pit during the Middle Eocene. Each forms a distinct component of the pollen assemblage within a small basin. A successional change in the local vegetation accounts for part of the variation in the lignite zone, a change in response to edaphic conditions. The assemblage does not represent a regional climax flora but a mixture of grains derived from several plant associations and successional change. These elements must be considered in estimating the structure of paleocommunities and past climates.

The vertical change of the pollen types and relative abundance caused by the influence of different source areas present limitations for stratigraphic correlation, based on pollen, among small, nonmarine, disjunct localities. The extent of palynomorph variation at one small site suggests that attempts to stratigraphically relate sites dispersed over the entire Mississippi Embayment, particularly in its nonmarine sediments, may not be practical. The amount of variation between the clay and lignite zones at Miller Clay Pit is as great as or greater than variation of some pollen types used to date Middle Eocene deposits in the upper portion of the Embayment (Tschudy 1973, Elsik 1974, Elsik and Dilcher 1974). The amount of microfossil variation found at Miller Clay Pit indicates a need for more extensive investigation into within-site pollen fluctuations.

REFERENCES

Berglund, B. E. 1973. Pollen dispersal and deposition in an area of southwestern Sweden--some preliminary results. In: H. J. B. Birks and R. G. West (eds.), Quaternary Plant Ecology, p. 117-129. Oxford: Blackwell Scientific Publications.

Berry E. W. 1914. The affinities and distribution of the lower Eocene floras of southeastern North America. Amer. Philos. Soc. Proc., 53: 129-250.

_____. 1916. The lower Eocene floras of southeastern North America. U.S. Geol. Surv. Prof. Paper 91, 481 p., 117 pls.

_____. 1924. The Middle and Upper Eocene floras of southeastern North America. U.S. Geol. Surv. Prof. Paper 92, 206 p., 65 pls.

_____. 1930. Revision of the Lower Eocene Wilcox flora of the southeastern United States. U.S. Geol. Surv. Prof. Paper 156, 196 p., 50 pls.

_____. 1937. Tertiary floras of eastern North America. Bot. Rev., 3: 31-46.

_____. 1941. Additions to the Wilcox flora from Kentucky and Texas. U.S. Geol. Surv. Prof. Paper 193E, p. 83099, 5 pls.

Crowder, A. A. & Cuddy, D. G. 1973. Pollen in a small river basin: Wilton Creek, Ontario. In: H. J. B. Birks and R. G. West (eds.), Quaternary Plant Ecology, p. 61-77. Oxford: Blackwell Scientific Publications.

Dilcher, D. L. 1971. A revision of the Eocene flora of southeastern North America. The Paleobot., 20: 7-18.

_____. 1973. A paleoclimatic interpretation of the Eocene floras of southeastern North America. In: A. Graham (ed.), Vegetation and Vegetational History of Northern Latin America, Ch. 2, p. 39-59. Amsterdam: Elsevier Publishing Co.

Elsik, W. C. 1974. Characteristic Eocene palynomorphs in the Gulf Coast, U.S.A. Palaeontographica, Abt. B, 149: 90-111.

_____ & Dilcher, D. L. 1974. Palynology and age of clays exposed in Lawrence Clay Pit, Henry County, Tennessee. Palaeontographica, Abt. B, 146: 65-87.

Flenley, J. R. 1973. The use of modern pollen rain samples in the vegetational history of tropical regions. In: H. J. B. Birks and R. G. West (eds.), Quaternary Plant Ecology, p. 131-141. Oxford: Blackwell Scientific Publications.

Janssen, C. R. 1966. Recent pollen spectra from the deciduous and coniferous-deciduous forests of northeastern Minnesota: a study in pollen dispersal. Ecology, 47: 804-825.

_____. 1973. Local and regional pollen deposition. In: H. J. B. Birks and R. G. West (eds.), Quaternary Plant Ecology, p. 31-42. Oxford: Blackwell Scientific Publications.

Lesquereux, L. 1859. On some fossil plants of recent formations. Amer. J. Sci. (2nd Ser.), 27: 359-366.

Muller, J. 1959. Palynology of recent Orinoco delta and shelf sediments: reports of the Orinoco shelf expedition, Vol. 5. Micropaleont., 5: 1-32.

Parks, W. S. 1971. Tertiary and Quaternary stratigraphy in Henry and northern Carroll Counties, Tennessee. J. of Tennessee Acad. Sci., 46: 57-62.

Potter, F. W., Jr. 1976. Investigations of angiosperms from the Eocene of southeastern North America: pollen assemblages from Miller Pit, Henry County, Tennessee. Palaeontographica, Abt. B, 157: 44-96.

Tauber, H. 1967. Investigations of the mode of pollen transfer in forested areas. Rev. Palaeobot. Palynol., 3: 277-286.

Tschudy, R. H. 1973. Stratigraphic distribution of significant Eocene palynomorphs of the Mississippi Embayment. U.S. Geol. Surv. Prof. Paper 743B, 24 p., 4 pls.

TOWARD AN UNDERSTANDING OF THE REPRODUCTIVE BIOLOGY OF FOSSIL

PLANTS

Thomas N. Taylor

Department of Botany

The Ohio State University 1735 Neil Ave., Columbus, OH

ABSTRACT

During the last several years there has been an increased emphasis in several areas of paleobotany that have focused on the reproductive biology of fossil plants. Several factors have played an important role in stimulating research along these lines. The present paper demonstrates several examples by which the reproductive parameters of fossil plants may be investigated. The total analysis of fossil plant reproductive systems not only provides an additional dimension for comparative studies with closely related extant forms, but moreover allows an opportunity to investigate and more accurately define heretofore abstract concepts concerning the evolution of certain reproductive systems.

INTRODUCTION

During recent years some paleobotanists have detailed features associated with the developmental phenomenon of fossil plants in what have been termed ontogenetic studies (Delevoryas, 1964, 1967). Most of these investigations have utilized structurally preserved fossil plants that have provided a wealth of information that has been used as another parameter by scholars of fossil botany in more precisely understanding the total biology of fossil organisms. These studies have not only provided information concerning various levels of growth and development, but have also contributed information that has been useful in more accurately reflecting a natural system of classification.

The understanding of some developmental parameters among certain

77

fossil plants has provided an opportunity to investigate various aspects of these plants that may conveniently be included under the heading of reproductive biology. Such studies have prompted paleobotanists not only to raise fundamental questions concerning the structure and function of various reproductive organs, but moreover to consider such aspects as the adaptive significance of certain structures, as well as strategies associated with reproduction, and the evolutionary implications involving various types of reproductive systems.

There are several factors that have played an important role in establishing a framework for studies directed at the reproductive biology of fossil plants. These include the utilization of more sophisticated techniques in specimen preparation and information extraction, the more widespread use of developmental parameters in specimen selection, and the consideration of more biologically oriented questions about fossil plants.

It is the intent of the present paper to discuss several examples that demonstrate methods of investigating the reproductive biology of fossil plants. Although examples might be cited from almost any geologic period or group of organisms, those delimited here include fossil plants of Pennsylvanian age that are preserved in calcium carbonate petrifactions known as coal balls. The reader is referred to the various literature citations within the body of the text for references to preparation techniques and procedures used with the various fossils.

LYCOPODS

The structural organization of the reproductive systems produced by some fossil organisms is uniquely suited to a study of reproductive mechanisms. Within the arborescent members of the Carboniferous order Lepidodendrales are taxa that produced sporophylls aggregated into loosely arranged strobili (Fig. 1). Several different cone organizational types are known from rocks of Carboniferous age, and include forms that produce only one type of spore (monosporangiate) as well as types that contain spores of two distinct sizes, and that presumably functioned differently in the biology of the organism. The large size of such cones, together with the sequential development of the parts, provides a convenient reproductive system that can be studied in detail. Because these cones matured in a sequential pattern the sporangial contents (spores) can be extracted from various levels and examined with reference to the stage of development reached. Using such material paleobotanists have a unique opportunity to extrapolate between such developmental stages, and by this procedure to piece together the probable structural events in the formation of various reproductive features.

Microspores

Microspores have been extracted from sporangia high in the
apex of a cone similar to that illustrated in Fig. 2. The spores
are still within the tetrad configuration, obviously having just
completed the meiotic process. Spores extracted from microsporangia
at a lower level of the cone (Fig. 3) have separated from the tetrad,
increased slightly in size, and may or may not display additional
ornamentation of the exine. Studies of this type not only provide
a framework for investigating the potential stages in the development of the microgametophyte phase of these plants, but further
assist in characterizing ontogenetic stages of spores from phases
that accurately reflect biological species. Such pollen and spore
studies are especially important in biostratigraphic considerations.

Megagametophytes

The larger arborescent lycopod spores, the megaspores, are
providing an increasingly clearer picture of several aspects of
fossil lycopod reproductive biology. Detailed studies like those
of Galtier (1964) and Brack (1970) have demonstrated the presence
of well preserved endosporic megagametophytes. In some forms
cellularized megagametophytes, some containing well preserved,
presumably absorptive rhizoids, are found (Fig. 5). In many of
these megagametophytes only the upper half of the megaspore is
cellularized (Fig. 4); the lower portion may represent a free
nuclear stage similar to that found within the megagametophytes
of several extant free sporing members of the Lycopodiophytina.

In several of these cellularized megagametophytes exquisitely
preserved sex organs, archegonia (Fig. 5), have been described
(Brack, 1975). By examining many megaspores from a single sporangium, as well as several sporangia from the same cone, it has been
possible to detail the stages in the development of the archegonium.
Figure 6 illustrates a polar view of an archegonium consisting of a
single tier of four neck cells. Brack-Hanes (1975) has demonstrated
that the mature archegonium of this fossil lycopod species consists
of four tiers of neck cells and an inflated venter containing the
egg. This study indicates that as egg maturity is reached the pro-
truding archegonial neck is broken off, thereby enhancing sperm
access. This structural feature is similar to the mechanism in
some extant lycopods. Not only do studies of this type provide information regarding some of the stages in the development of the
mature gametophyte prior to fertilization, but also underscore the
value of the gametophyte as useful in both systematic and phylogenetic considerations.

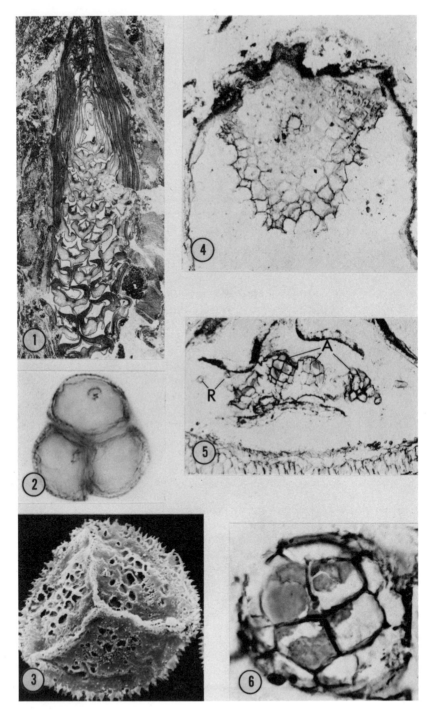

Megaspore Wall Ultrastructure

The recent examination of the wall of lycopod megaspores via transmission and scanning electron microscopy not only underscores the importance of such instrumentation in studies considering reproductive mechanisms of fossil plants, but in addition constitutes a level of information that has been heretofore unobtainable. By extracting megaspores from a Lepidocarpon cone it has been possible to determine information relative to the changes in the organization and ultrastructure of the wall during the development of the megagametophyte or embryo. Ultrastructurally the sporopollenin units that constitute the megaspore wall consist of bead-like units that are fused together (Fig. 9). This type of organization is present in megaspores extracted from relatively high in the tip of the cone. Mature megaspores extracted from lower regions of the same cone may be up to 4.0 mm long. When spores of this latter type are macerated from more basal regions of Lepidocarpon cones the ultrastructure of the wall is similar to that illustrated in Fig. 10. This unique structural property of the megaspore wall that provides for rapid expansion also provides a continuing source of protection for the developing megagametophyte or embryo inside.

←————

Fig. 1. Longitudinal section of Lepidocarpon cone showing apical, immature sporangia. X 5.5. - Fig. 2. Partial tetrad of Lycospora microspores. X 1200. - Fig. 3. Proximal surface of mature Lycospora microspore. X 2400. - Fig. 4. Section of lycopod megaspore showing cellular nature of female gametophyte. X 210. - Fig. 5. Oblique longitudinal section of megaspore apex showing several archegonia (A) and rhizoid remnants (R). X 200. - Fig. 6. Transverse section of archegonium showing position of tiered neck cells. X 1000.

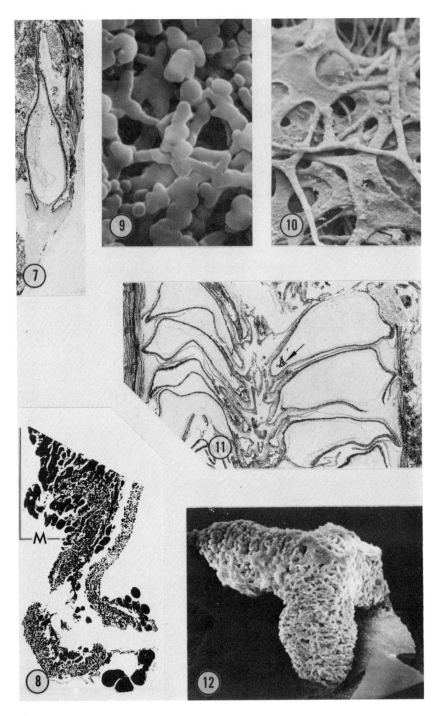

Recent studies of other lycopod megaspores suggest another parameter that may be explored in discussions about the reproductive biology of these organisms. Figure 12 shows a large sporopollenin unit (massa) that covers the proximal surface suture of the megaspores of another lycopod species. This unusual feature is found on the megaspores in certain lycopod cones that contain sporangia that lack enveloping protective structures termed lateral laminae (Fig. 7). In these cones the megaspores with the massa are oriented so that the proximal surface suture is closest to the axis of the cone (Fig. 11). This is in direct contrast to the usual situation of a cone type like Lepidocarpon in which the megaspore is oriented so that the proximal surface suture is directed distally within the sporangium, or furthest from the cone axis. Moreover, the mega-sporangia of this latter cone type possess   enveloping lateral lam-inae (Fig. 1) and contain megaspores that lack a distinct proximal surface massa. It is suggested that the adaptive value of the massa may represent protection for the developing megagametophyte within the sporoderm of the megaspore. Moreover, the ultrastructure of the massa (Fig. 8) suggests a greatly increased surface area that may have functioned as a moisture retaining mechanism. If opera-tive in this way the massa may have also aided in providing channels for presumably motile gametes to reach the archegonial necks of the megagametophyte.

Fig. 7. Longitudinal section of Achlamydocarpon megasporangium showing reduced lateral laminae. X 4. - Fig. 8. Ultrathin section of Achlamydocarpon megaspore showing nature of massa (M) and detail of spore wall. X 800. - Fig. 9. Sporopollenin units of immature megaspore wall extracted from apical sporangium illustrated in Fig. 1 cone. X 7000. - Fig. 10. Sporopollenin units of mature megaspore wall. X 1000. - Fig. 11. Longitudinal section of Achlamydocarpon cone showing proximal position of functional mega-spore suture (arrow). X 8.5. - Fig. 12. Massa covering functional megaspore suture. Compare with massa in Fig. 8. X 180.

SEED FERNS

Tapetal Remains

Recent studies (Taylor, 1976) indicate the sporangial contents exclusive of the spores may constitute an important information source regarding the reproductive biology of some fossil plant groups. It is now apparent that with the aid of transmission electron microscopy information regarding the type of tapetum produced within the sporangium of fossil plants is readily available and potentially important.

At some stage in the development of sporogenous tissue in vascular plants, cells lining the internal surface of the sporangial cavity break down to form a nutritive material termed the tapetum that is associated with the developing spores. Two principal types of tapetum are recognized based principally on the manner in which they function during sporogenesis. Of these two types, the secretory form of tapetum has been extensively studied. One of the characteristics of the secretory tapetum is the production of sporopollenin particles on the inner surface of the sporangial walls during microsporogenesis. These particles have been variously termed orbicules, plaques, or ubisch bodies. Typically they consist of spherical structures only a few micrometers in diameter. The wall is thick and typically displays an ornamentation pattern similar to that of the mature spore wall. In many instances ubisch bodies are often associated with a series of delicate membranes that have been termed peritapetal membranes or tapetal membranes.

The ultrastructural examination of particles on the surfaces of certain Pennsylvanian age seed fern pollen grains of the Schopfipollenites-type has revealed the presence of ubisch bodies and associated tapetal membranes (Fig. 13). Orbicules associated with these fossil pollen grains characteristically measure less than a micrometer in diameter and appear to be hollow. The external surface of the wall is generally smooth (Fig. 13). Associated with these structures are a series of structurally complex membranes that appear to be morphologically identical to the tapetal membranes of certain extant plants (Fig. 17). To date the exact function of ubisch bodies and tapetal membranes is not known. Some have suggested that ubisch bodies constitute the sites of sporopollenin depolymerization, while others ascribe a function associated with sporopollenin transfer. Other potential functions that have been suggested include a method of pollen dispersal. In a similar manner the exact function of the tapetal membranes still remains highly speculative. One often cited function includes protection for the developing pollen mass.

Although the discovery of ubisch bodies associated with the pollen grains of Pennsylvanian age seed plants does not markedly contribute to our understanding of the function of these structures, the realization that such units were produced strongly suggests that the secretory type of tapetum was established in at least one group of early seed plants during Pennsylvanian time, and that this type may represent the primitive tapetal system. Continued studies directed at the type of tapetum present in Carboniferous age vascular plants may not only fill in the informational gap with reference to the way in which such a tissue system functioned in the production of spores and pollen grains within fossil groups, but may ultimately provide information about the phylogeny and subsequent evolution of the tapetum in all vascular plants.

Ovules

The ovules of some Paleozoic seed plants appear to have functioned similarly to those of certain extant gymnospermous types. In several fossil forms there is ample evidence that at some stage in the development of the ovule a sticky exudate was produced from the orifice of the micropyle and apparently functioned as a mechanism for trapping pollen grains. As in extant gymnospermous ovules, the pollen grains were pulled toward the pollen chamber as these pollination droplets dried out. Evidence suggests that once the grains were within the pollen chamber the beak of the nucellus may have closed thereby providing a sealed chamber for the subsequent development of the gametophyte phase.

Pollen

Information is now beginning to accumulate regarding the disposition of seed plant pollen grains once they were shed from the pollen sac. Although information about the pollination agent of many of these ancient seed plants is unknown, several of the steps involved in the subsequent development and fate of the grains is slowly emerging.

The occurrence of pollen grains of the same type within the pollen chambers of fossil ovules not only is important in demonstrating the biological relationship between isolated plant parts, but moreover, provides an obvious reference point from which to base subsequent studies concerning the development of the microgametophyte phase.

Pollen Tubes

Until recently information concerning the presence of pollen tubes associated with Paleozoic ovules was only speculative. In 1972 Rothwell demonstrated the existence of a branched pollen tube

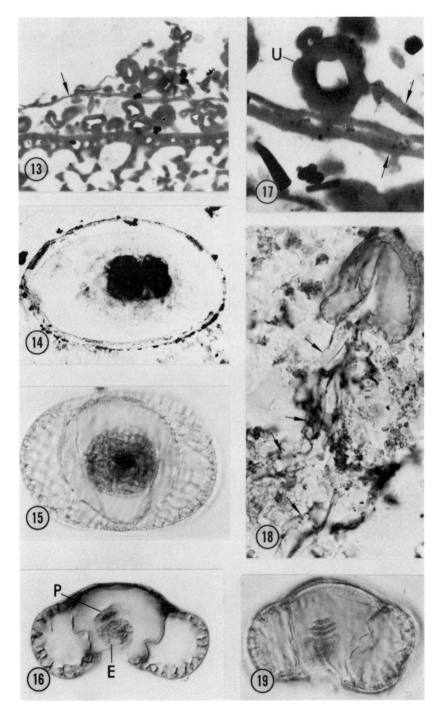

Fig. 13.   Section of Pennsylvanian pollen grain wall showing
spherical ubisch bodies and tapetal membrane remnants (arrow). X
9550. - Fig. 14.   Section of Schopfipollenites grain with two dense
structures suggestive of sperm. X 154. - Fig. 15.   Polar view of
seed fern pollen grain showing nuclei of microgametophyte. X 1500. -
Fig. 16.   Two-celled microgametophyte of Pennsylvanian seed fern.
E=embryonal cell, P=primary prothallial cell. X 1500. - Fig. 17.
Ubisch body (U) and complex nature of tapetal membrane (between
arrows). X 37,000. - Fig. 18.   Seed fern pollen grain with distally
germinating pollen tube (arrows). X 1300. - Fig. 19.   Four-celled
stage of seed fern microgametophyte. X 1500.

extending from the distal surface of a saccate pollen grain produced by a member of the seed fern family Callistophytaceae (Fig.
18). Of particular importance is the fact that the pollen grain
was observed within the nucellus of a Paleozoic ovule and apparently well along in the development of the microgametophyte. The
pollen tube extends from the distal surface and divides to form
three slender branches. The two longer branches are approximately
70μm long, while the shorter one measures 21μm long; pollen tube
diameter ranges from 3-7 micrometers. Although it can not be
determined whether the pollen tube was truly siphonogamous or merely
haustorial in function, the discovery is significant in demonstrating that pollen tubes germinating from the distal surface of pollen
grains had evolved by Middle Pennsylvanian time. This information,
together with the newly acquired information about the organization
of the microgametophyte of morphologically identical pollen grains,
indicates that the microgametophytes of at least one group of
Paleozoic seed plants were relatively advanced.

Microgametophytes

One of the most significant discoveries in paleobotany during
the last several years is the report of stages in microgametophyte
development within a group of Pennsylvanian age seed plants (Millay
and Eggert, 1974). From pollen grains extracted from synangia
these authors have been able to demonstrate several stages in the
development of the microgametophyte of this monosaccate pollen
grain type. Figure 16 illustrates a lateral view of a two celled
stage showing the larger embryonal cell and smaller primary
prothallial cell. The three-celled stage is illustrated in Figure
20. In Fig. 19 the three prothallial cells of the four-celled
stage are of approximate equal size. A polar view of the same
grain (Fig. 15) shows a well preserved nucleus of one cell of an
inverted four-celled stage of the microgametophyte. What is most
remarkable about the microgametophytes described by these authors
is the morphological similarity to the gametophytes of certain
extant members of the Coniferales. This discovery indicates that
by Pennsylvanian time the morphological organization of seed
plant microgametophytes was well established, and further provides
evidence that contradicts long held beliefs that the microgametophyte of seed plants has evolved via the reduction and loss of
almost all vegetative or sterile cells of the antheridium.

Additional evidence detailing the microgametophyte phase in
Paleozoic seed plants is the remarkable preservation illustrated
by Stewart (1954) of a pollen grain of another Paleozoic age seed
fern (Medullosaceae). Within the pollen chamber of Pachytesta
hexangulata is a pollen grain (Fig. 14) that contains two dense
ovoid structures that are morphologically similar to the sperm
characteristic of living members of the Cycadales. These structures

measure approximately 65 X 85 μm in diameter and morphologically appear similar to the sperm produced by the extant genus <u>Microcycas</u>. Substantiating the fact that the microgametophyte within this ovule was developmentally mature at the time of fossilization, is the presence of a partially cellularized megagametophyte, containing a structure that both topographically and morphologically resembles an archegonium. The dense structure contained within the presumed venter of the archegonium is thought to represent the remnants of the egg. Like the microgametophytes described from the Callistophytaceae, those of the Medullosaceae appear to be morphologically identical to extant forms thought to be closely related.

A further example of the source of information that may be extracted from fossil plants relative to the reproductive biology of the organism is illustrated by the recent study of <u>in situ</u> microspores (Taylor and Millay, 1976). These authors have described structures contained within the pollen grains of a cone type possibly representing either a coniferophyte or cycadophyte strobilus. The microspores of this cone consistently contained dense circular structures that are considered to constitute the shrunken remains of microspore cytoplasm and nuclei (Fig. 21, 22). None of the grains contained structures that could be interpreted as representing prothallial cells. The absence of such cells from the microspores of this cone type at the time of sporangial dehiscence suggests that prothallial cells were never produced. Among extant plants the absence of prothallial cell production during microgametophyte ontogeny is regarded as an advanced feature in some groups. The fact that prothallial cells are absent in these grains serves to indicate the level of evolution of the reproductive system of at least one group of Pennsylvanian age pollen producing plants.

CONIFERS

Embryos

Thus far in this paper the discussion has centered around features normally associated with the gametophyte phase of the reproductive system of fossil plants. To date relatively little is known about the immediate post fertilization stages in Paleozoic age plants. Despite the fact that numerous structurally preserved ovules of Paleozoic age have been described, there is only a single account in which an embryo has been described and illustrated (Brown and Miller, 1973). The recent report by Phillips, Avcin and Schopf (1975) in which embryos and sporelings have been reported as occurring in the genus <u>Lepidocarpon</u> is encouraging support that information on this phase in the reproductive biology of fossil plants will be forthcoming.

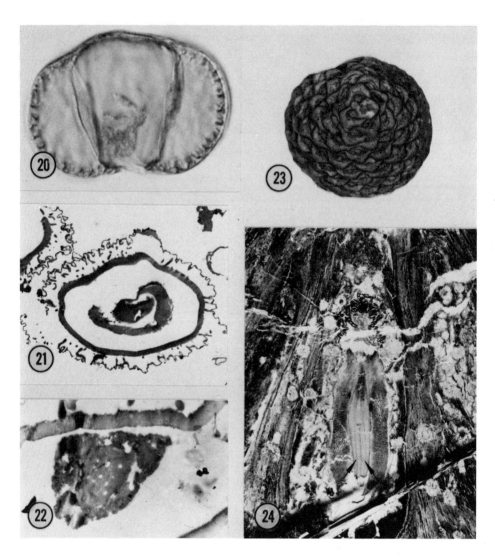

Fig. 20. Lateral view showing three-celled stage of microgametophyte. X 1500. - Fig. 21. Lasiostrobus microspore showing
nuclear and cytoplasmic contents. X 2400. - Fig. 22. Section of
Lasiostrobus microspore showing differential density of contents.
X 10,980. - Fig. 23. Apical view of silicified araucarian cone
showing spirally arranged ovuliferous scales. X 1. - Fig. 24. Section of araucarian seed containing embryo embedded in megagametophyte. Arrows indicate the position of the vascularized cotyledons.
X 10.

Despite the paucity of remains of embryos in Paleozoic age sediments there are several reports of embryos contained within the seeds of Mesozoic age seed plants. Recently, Stockey (1975) has described several esquisitely preserved silicified ovuliferous cones from the Cerro Cuadrado Petrified Forest of Patagonia (Fig. 23). Thin sections of the silicified cones reveal the presence of spirally arranged seeds embedded in the upper surface of the ovuliferous scale. Contained within some seeds are well preserved embryos (Fig. 24). Longitudinal sections of the embryo show the typical features of a conifer embryo including shoot apex, cotyledons, and root meristem initials. Depending upon the stage of development prior to fossilization some embryos show features of differentiating vascular tissue. Of particular interest in this study is the information that is presented regarding the similarity of these Mesozoic age embryos to those of certain extant conifers. Such studies will not only contribute markedly toward a more accurate picture of the embryology of these fossil forms, but more importantly will provide a basis for considering the evolution of such features as polycotyledony and polyembryony in coniferous taxa.

CONCLUSION

The preceding represents a few examples that serve to underscore the potential value of studies directed at the reproductive mechanisms and biology of fossil plants. Similar examples and approaches may be found among additional groups of fossil taxa encompassing almost all periods of geologic time. Despite the apparent limitations imposed by the nature of the organisms under investigation, studies of fossil plants are currently utilizing techniques and methods of inquiry that constitute some of the most sophisticated presently known within the biological sciences. One need only cite the recent study by Niklas (1976) in which parameters of chemotaxonomy are utilized in the study of fossil organisms, or the report by Schmid et. al. (1976) in which immunochemistry is utilized in order to realize the increasing research methodologies being applied in paleobotany.

Studies that consider the broad ramifications of reproductive biology utilizing fossil plants will have to be viewed as a continuum of biological systems. Paleobotanists will, by necessity, have to cross traditional boundaries in attempting to devise new methods of extracting information from fossil plants that concern reproductive mechanisms. As information about reproductive mechanisms becomes more accessible it should be added to anatomical and morphological data that can be used in systematic and phylogenetic considerations. Such features as the number of archegonial neck cells, cotyledons, position and number of prothallial cells etc. will not only provide additional information about the total biology of the organism, but may be useful in more accurately understanding the evolution of such features. Lastly, information associated with the reproductive biology of the plant provides a direct means whereby the potential adaptive significance of certain features and reproductive mechanisms may be properly interpreted.

ACKNOWLEDGEMENTS

The author is appreciative to the following individuals for providing permission to use some of the figures included within this paper: Dr. Sheila Hanes, Dr. Michael A. Millay, Dr. Gar W. Rothwell, Dr. Wilson N. Stewart, and Ms. Ruth A. Stockey.

This study was supported in part by the National Science Foundation (BMS74-21105).

LITERATURE CITED

Brack, S. D.1970. On a new structurally preserved arborescent lycopsid fructification from the Lower Pennsylvanian of North America. Amer. J. Bot. 57:317-330.

Delevoryas, T. 1964. Ontogenetic studies of fossil plants. Phytomorphology 14:299-314.

_____. 1967. Further remarks on the ontogeny of certain Carboniferous plants. Phytomorphology 17:330-336.

Galtier, J. 1964. Sur le gametophyte femelle des Lepidodendracees. C. R. Acad. Sci. 258:2625-2628.

Hanes, S. D. 1975. Structurally preserved Lepidodendracean cones from the Pennsylvanian of North America. Ph.D. Dissertation, Ohio University, 112 p.

Millay, M. A. and D. A. Eggert. 1974. Microgametophyte development in the Paleozoic seed fern family Callistophytaceae. Amer. J. Bot. 61:1067-1075.

Miller, C. N. and J. T. Brown. 1972. Paleozoic seeds with embryos. Science 179:184-185.

Niklas, K. J. 1976. Chemical examination of some non-vascular Paleozoic plants. Brittonia 28:113-137.

Phillips, T. L., M. J. Avcin, and J. M. Schopf. 1975. Gametophyte and young sporophyte development in Lepidocarpon. Bot. Soc. Amer. Abstr., p. 23 Lawrence, Kansas.

Rothwell, G. W.   1972.   Evidence of pollen tubes in Paleozoic
    pteridosperms.   Science 175:772-774.

Schmid, R. S., M.Wolniak, and V. J. Vreeland.   1976.   Electron
    microscopy and immunochemistry of Prototaxites.   Bot.
    Soc. Amer. Abstr., p. 31, Lawrence, Kansas.

Stewart, W. N.   1954.   A new Pachytesta from the Berryville
    locality of southeastern Illinois.   Amer. Midl. Nat. 46:
    717-742.

Stockey, R. A.   1975.   Seeds and embryos of Araucaria mirabilis.
    Amer. J. Bot. 62:856-868.

Taylor, T. N.   1976.   Fossil ubisch bodies.   Trans. Amer. Micros.
    Soc. 95:133-136.

_____ and M. A. Millay (in press).   The ultrastructure and
    reproductive significance of Lasiostrobus microspores.   Rev.
    Palaeobot. Palynol.

A GEOBOTANICAL OVERVIEW OF THE BRYOPHYTA

Harvey A. Miller

Florida Technological University

Orlando, Florida 32816

ABSTRACT: Macroevolution of mosses is considered from view-points of the fossil record, relative orientations of continents, paleoclimatology, morphology, and distributional phenomena. Incomplete evidence indicates: 1) mosses are offshoots of early rhyniophyte and zosterophyllophyte stock;   2) Sphagnales were set off well before the end of the Permian;   3) Bryales, in the broadest sense, were diverse and widely distributed by Permian time in mesic to wet, temperate sites; 4) diversity among all groups of bryophytes was considerably reduced by the Permo-Triassic desert episodes on most major land masses;   5) the present-day moss generic flora has changed little since the close of Cretaceous time and has a preponderance of drought-resistant groups some of which extend back to the Paleozoic; and 6) present day diversity reflects radiate evolution by survivors of the desert episodes into new niches created by the ascendancy of angiospermous vegetation.

Mosses   are   among   the   oldest groups of extant terrestrial plants with the result that their early evolutionary history has been obscured largely by the limited fossil evidence so far reco-vered.   Recently, as new techniques have been developed for hand-ling delicate fossils, the number of pre-Tertiary bryophytes has increased markedly (e.g., Brown and Robison, 1974).   Although the record is yet, and must remain, imperfectly known, some extremely important benchmarks have been established in the long history of mosses.   Further discoveries will surely further close gaps and clarify seeming evolutionary anomalies.

In two previous publications on macroevolution in bryophytes (Miller, 1974 a,b), I outlined the sequence of events immediately following, or associated with plant transmigration to the land as indicated by a rather good,and fast improving, known fossil record dating from the Middle Ludlovian strata of the Upper Silurian containing Cooksonia onward through the Devonian (Banks, 1968a). The early Devonian, thus, was an appropriate time for elaboration of bryophytic plants and so selected elements of that ancient flora are reviewed as they seem to relate to the possible origin of mosses.

If we turn our attention briefly to some of the earliest known terrestrial forms and their morphologies we see that Rhyniophyta (Bold, 1973) were diminutive, dichotomously branched plants of wide geographic distribution in Upper Gedinnian time. Representative genera are Rhynia, Eogaspesia, Taeniocrada, and the oldest known land plant of all, Cooksonia, which dates from the Upper Silurian.

If we accept that Cooksonia, or similar rhyniophyte, is representative of the earliest land plants, we must seek the origin of mosses at about the same time and at the same evolutionary level or in subsequently evolved groups.

The naked stems of the Rhyniophyta are certainly not moss-like but sporangia are terminal and may dehisce longitudinally in a manner reminiscent of Andreaeales. The first instances of enations arise among some of the Zosterophyllophyta which are characterized overall by lateral sporangia with dehiscence along the distal edge. Banks (1968 a,b) has suggested that Zosterophyllophyta similar to Sawdonia may provide the basic stock for Asteroxylon which he viewed as a primitive lycophyte from which several lines evolved. Leaf traces in Asteroxylon are weak and do not penetrate the appendages, but are typical of leaf traces otherwise.

Diversity known within the Lower Devonian flora shows clearly both a considerable range of genetic potential among the several taxa and some possible combinations of characteristics. If we take a morphogenetic view of these variations and possible combinations not yet discovered and accept the strong probability that vegetative portions could have been either haploid or diploid (Merker, 1958; Lemoigne, 1968) while phenotypically distinguishable only when reproductive structures are present, then the gene pool contains the basic building blocks for the mosses. This is not to say that the mosses are derived from lycopods but simply that the evolutionary continua for terrestrial plants which can be read partially from the fossil record indicate that basic structural features incorporated into both mosses and lycopods were present during the period under consideration. The mosses further incorporate a combination of morphological expression in the monoploid-diploid alternation of generations which achieves a greater level of diversity, though not the independence, known for a spectrum of diphasic algal groups sometimes proposed as ancestral in some manner to the bryophytes. Despite the existence of diphasic algal groups, present evidence is overwhelmingly against algae being directly ancestral to any bryophytic plants (Steere, 1969; Miller, 1974a).

Another evolutionary development that bears on the problem is that heterospory did not originate until the Middle Devonian. This means that a 25 million year period was available for diversification of homosporous plants, including bryophytes, prior to the existence of heterosporous groups. Thus, the time of origin of heterospory may be an important, albeit indirect, marker in determining a minimum age for bryophytic lines.

The central strand in the rhyniophytes in all cases known is terete with central protoxylem (fig. 1) (Banks, 1968a). Zosterophyllophyta appear in the fossil record first in Gedinnian-Siegenian strata as species of Zosterophyllaceae (Zosterophyllum) from Australia, Belgium, Wales, and Scotland and as a second family, Gosslingiaceae, present in Siegenian strata. The central strand in Gosslingia (fig.1) is somewhat larger than in Rhynia and Hostinella and is an elliptic, exarch protostele. Further, the cortical region is composed of thick-walled cells. Both Crenaticaulis and Sawdonia in the same family have terete, exarch protosteles (Banks and Davis, 1969; Hueber, 1971) not unlike those of

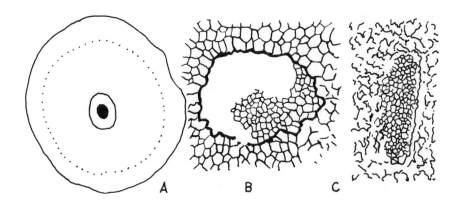

Figure 1. A. Cooksonia stem cross-section; B. Detail of central centrarch protostele in Cooksonia; C. Gosslingia cross-section showing exarch elliptic protostele. (Based upon photographs in Banks, 1968a)

some mosses (Kawai, 1971 a,b,c; Hébant, 1964) as seen in cross-section. Sufficient stelar variation has been revealed by recent discoveries of early Devonian plants, some yet not formally reported, that it seems clear that bryophytic "central strands" fall well within the range of structure acceptable to botanists as vascular tissue. Further, the variations in cortical structure known from the fossils can also be found among mosses (e.g., Kawai, 1971a). The differences between the "vascular plants" and bryophytes pale markedly in light of these discoveries (e.g., Finocchio, 1967; Hébant, 1964, 1967; Hébant and Schafer, 1974).

Until recently, Sawdonia (as Psilophyton ornatum) was the single multicellular-spined zosterophyllophyte. To this can now be added Crenaticaulis and Kaulangiophyton which also bear multi-cellular enations (Banks and Davis, 1969; Gensel, Kasper, and Andrews, 1969). Expansion and flattening of these surface enations is a minor evolutionary change. We have already noted the presence of leaf traces in Asteroxylon which is known to have existed and to have been at an evolutionary level comparable in several respects to mosses during late Lower Devonian.

Vegetative axes of Rhyniophyta and Zosterophyllophyta provide relatively non-controversial evidence of an organizational level not markedly different from vegetative portions of bryophytes with the zosterophyllophytes having greater parallels with the mosses. The case is less clear, however, with regard to gametangia. Lemoigne (1970) has described and figured structures represented as archegonia and antheridia on the prostrate axis of Rhynia. Investigations are presently being undertaken by others to examine these structures further. If the work is borne out, but it is yet considered controversial by many paleobotanists, then Merker's (1958) rather remarkable interpretations will be confirmed that Rhynia is comprised of a prostrate rhizoid bearing gametophyte with the erect sporophyte arising from an embedded haustorium. Such a situation would parallel the bryophytes very closely and indicate that the present pattern of alternation of generations characterized by some continuing dependence of the sporophyte upon the gametophyte follows an ancient pattern indeed.

Sporangia of rhyniophytes and zosterophyllophytes differ somewhat in the pattern of dehiscence with the rhyniophytes opening longitudinally in the manner of many hepatics as opposed to transverse dehiscence observed in zosterophyllophytes and most mosses. Horneophyton, a rhyniophyte, had a well developed columella, as perhaps did others.

Spores of ancient land plants are uniformly trilete, as in bryophytes, and within a proper size range for mosses and liverworts. In general, spore surfaces for the oldest land plants were simply ornamented, if at all, and the later derived groups still within the Devonian frequently have more elaborately sculptured spores. Much remains to be learned from Devonian spores, but nothing so far known would disqualify an early Devonian origin for several "bryophytic" lines.

What we see in the rhyniophytes aand zosterophyllophytes are not bryophytes but a constellation of organisms which possess many attributes in common with one or more of the several recognized bryophytic lines. The key to success of terrestrial plants is undeniably associated with internal structural adaptations which accomplish translocation of essential materials, especially water and minerals. Evolutionary experiments with vascularization were numerous in the early Devonian with various patterns of cellular maturation in conducting tissues. Plants of small stature adapted to mostly moist, and perhaps continuously shaded, environments characteristic of bryophytes were not subject to intense selective pressures acting upon other components of the vegetation. Those forms which may have evolved more elaborate vascular systems and outgrew the comparatively uncompetitive microhabitat apparently did not survive in recognizable form and left behind "nongenerate" vascular plants--i.e. bryophytes.[1]

Vascularization in these plants exists, as it always has, at the early unspecialized rhyniophyte-zosterophyllophyte level. The bryophytes are not degenerate vascular plants as sometimes proposed,but vascularized plants in their own right whose true nature was obscured philosophically by the fast-fading (Steere, 1969; Miller,1974) morphological dogma that gametophytes, by definition, cannot be vascularized.

So far, no mosses in the strict sense have been turned up prior to later Carboniferous time and the preservation of the Carboniferous specimens is such that cellular details have remained unknown. Thus, their absolute identity to mosses, while highly probable, is not presently confirmable. The oldest unquestionable mosses so far reported appear in Permian deposits in Angaraland in Neuberg's now classic study on that rich fossil bryoflora. Unquestionably moss fossils have been found also among the Permian Glossopteris flora of Antarctica (Schopf and Miller, in prep.).

The presence of a variety of true mosses in widely separated regions in Permian deposits indicates that considerable diversity existed by that time. Further, it demonstrates that the bryoid habit was well established and had been for a long time. At least it seems unlikely that dispersal occurred across the Tethys between Angaraland and the Antarctic shield in the Permian (Schopf

---

[1] I am indebted to Dr. J. M. Schopf for his suggestion of the term "nongenerate" to describe bryophytic vascularization.

1970). If we allow that some connection, or perhaps better, a climatic continuity combined with land masses in proximity must have existed between Antarctica and Angaraland, we find ourselves looking at Schwarzbach's (1961) map of Devonian climates which places Australia, eastern China, most of Siberia, northern Europe and northern North America all at middle latitudes in the same (mostly northern) hemisphere. South America and Africa are in the opposite (mostly southern) hemisphere. This configuration provides a possibility for wide dispersal of ancient bryophytic stock. The glacier influenced Carboniferous Gondwana cool and moist climates would appear to be ideal for a flourishing bryoflora. Conditions were thus quite amenable for development of an extensive bryoflora and bryovegetation in pre-Permian times (fig. 2).

The end of the Paleozoic was marked by formation of deserts or mediterranean climates on land masses of the northern hemisphere while moist temperate climates were apparently closely restricted to a limited area of Gondwana now mostly part of Antarctica (King, 1961). The selective pressure on the Permian moss flora of northern lands favored those forms somehow adapted to survival through drought conditions.

If we consider those groups of mosses limited to a northern distribution (Table 1) we find included those adapted to deep shade and rock substrate (e.g., small cave conditions) in the Bryoxiphiales, Schistostegales, and Tetraphidales. Also, Fontinalales are aquatics which have the ability to persist in shaded intermittent rocky streams which probably existed in somewhat moister uplands. Buxbaumiales, a couple of which are known from Australia, can persist in gymnospermous forests under acidic edaphic conditions and are quite seasonal in their appearance. Only the Timmiineae do not fit the pattern well but they may be misplaced among northern groups. They were excluded from high latitude groups because I am aware of no <u>Timmia</u> in high southern latitudes.

The southern groups all extend into the Austral-Asian tropical mountains and in some ways imbricate with the tropical groups. We note, however, that in each of these southern groups, the taxa are characterized today by shade tolerant, hygrophilic types which would have thrived in the coolest mesic refugia in Gondwana during northern desert episodes. Numerous genera in several widely-distributed orders are restricted to the southern hemisphere with the result that the extent of southern floristic element is underrepresented in Table 1. Even so, a southern element can be loosely recognized at the ordinal-subordinal level.

The groups lumped under high latitude and altitude are comprised of many genera and species which exist under conditions of exposure and periods of moisture stress parallel to conditions that were avoided or adapted to in some manner by the small,

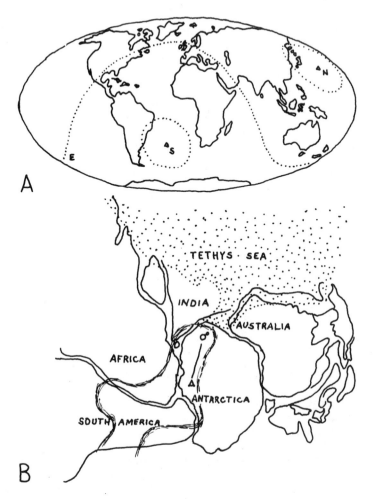

Figure 2. A. Climatic zones of the Devonian as suggested by relative locations of the equator (E), the position of the south pole (S), the north pole (N), and polar circles. (Based upon Schwarzbach, 1961) B. A tentative reconstruction of Gondwana continents in the Permian with the present position of the south pole indicated for reference. The *Glossopteris* flora, which included mosses, covered most of this area and the portion limited by the sketch line was a postglacial lake area. (Based upon Schopf, 1970)

Table 1.   Generalized Distribution of Orders of Mosses.

Northern Groups

Protosphagnales (Permian fossils from Angara)
Bryoxiphiales (1 genus and 2 species)
Schistostegales (monotypic-the "goblin gold" moss)
Bryales--Timmiineae (1 genus and 8 species)
Fontinalales (3 families, 5 genera, 32 species)
Tetraphidales (2 genera, 3 species)
Buxbaumiales (2 families, 3 genera, 23 species)

Southern Groups

Bryales--Rhizogoniineae (6 families, 13 genera-but only 1 with
     more than 10 species-63 species
Hookeriales--Ephemeropsidineae (1 genus, 2 species)
Hookeriales--Hypopterygiineae (2 families, 5 genera, 98 species
     with 80 in only 2 genera)
Dawsoniales (1 genus, 9 species)

High Latitude and Altitude Groups (many bipolar taxa)

Andreaeales (1-3 genera, ca. 100 species)
Encalyptales (2 genera, one monotypic, 34 species)
Bryales--Bartramiineae (6 genera, 25 species plus the widely
     distributed Bartramiaceae of 9 genera and ca.430 species)

Wet Tropical Groups (Principally)

Syrrhopodontales (7 genera, ca. 600 species)
Hypnodendrales (2 families, 6 genera, ca. 65 species)
Leucodontales--Racopilineae (2 families, 3 genera, 62 species)
Leucodontales--Neckerineae (4 families,35 genera,ca.500 species)
Hookeriales--Hookeriineae (6 families,45 genera,ca.780 species)

Widely Distributed Groups

Sphagnales (1 genus, ca. 200 species)
Archidiales (1 genus, 34 species)
Dicranales (8 families, 90 genera, ca. 2100 species)
Fissidentales (8 genera, ca. 1000 species)
Pottiales (2 families, 76 genera, ca. 1600 species)
Grimmiales (7 genera, ca. 430 species)
Orthotrichales (4 families, 28 genera, ca. 1050 species)
Funariales (7 families, 28 genera, ca. 460 species)
Bryales--Bryineae (3 families, 33 genera, ca. 1420 species)
Leucodontales--Leucodontineae(15 families,93 genera,ca.870 sp.)
Hypnobryales (11 families, 186 genera, ca. 3210 species)
Polytrichales (20 genera, ca. 380 species)

strictly holarctic groups. Their ability to survive under alpine conditions with (relict?) populations holding on, as if on islands scattered over vast distances could account for their present day distributions which parallel those of some alpine angiosperms.

It seems quite probable that the wet tropical groups were able to hang on in the warmer parts of Gondwana refugium with oceanic climates which could explain their considerable representation in the southern hemisphere. Numerous representatives of the groups listed are low altitude species or those which may survive wet-dry seasonal or monsoon climates as well as many of continuously moist sites.

The widely distributed groups are also those with the greatest numbers of genera and species. Included here are both rather genetically plastic and seemingly genetically static taxa. Their wide distribution at higher taxonomic levels with more limited ranges for lesser levels suggests strongly that the groups are of great antiquity and that geographic isolation has also been long operative. Sphagnales are plants which may survive wherever there is surface water and opportunity for development of a water-conserving polster, as for some *Sphagna* in Florida which seem remarkably drought resistant. The extant genus, *Sphagnum*, had relatives living in Permian Angara and it may itself have been present at that time. Its survival to the present over a broad geographic range is indicative of its success in colonizing suitable edaphic sites almost regardless of macro-climatic factors.

Archidiales are unique for their extraordinarily large spores, over 100 microns, which seem tolerant of prolonged drought. Many are quite ephemeral and are adapted to survival in a seasonally dry climate.

Polytrichalian vascularization, complete with leaf gaps and lignin, suggests an origin from a somewhat different gene pool than the Bryidae. This seems substantially confirmed by structure of the nematodont peristome and a base chromosome number of 7 vs. 10-14 for most Bryidae (Khanna, 1964). Several genera are rather local and with few taxa which may indicate a long isolation from parent stock. Representatives are found across a spectrum of climates and edaphic conditions though acidic, mineral-poor soils are quite acceptable substrates for many Polytrichales.

The remaining orders, all Bryidae, share several characteristics: all have large numbers of species; all are widely distributed, but with some localization of lesser taxonomic levels; and each has a drought survival strategy strongly represented among extant taxa. Thus, the present success of these groups has hinged upon a genetic plasticity which allowed for adaptations to a wide variety of habitats from seasonally dry climates and inhospitable cover vegetation to the plush vegetation and constant moisture of the cloud forest. Even though generalizations are often danger- ous out of context, I have attempted to point out features which exemplify some of the survival strategies.

Dicranales are often plants of exposed mineral soils although many epiphytic forms are known.   Some even appear on pine bark- generally a most inhospitable and acid substrate.   Many smaller and terrestrial species fruit profusely in contrast to larger, often epiphytic, forms which produce great numbers of leaf frag- ments, all with totipotent cells capable of producing new proto- nemata and plants.   In some families, specialized leaf fragments, propagula, are formed on the costa as in the Leucobryaceae and Calymperaceae. All are acrocarpous and most have a strong percur- rent to excurrent costa.

Fissidentales are mostly small mosses of damp shaded rocks although they may also be on tree bases or, in the case of larger species, terrestrial.   The distichous leaf arrangement provides a maximum exposure of leaf surface to light and the bulging to once- conic papillate cell walls so common in the group may serve to concentrate light and thus increase photosynthetic efficiency in their deeply shaded niche. The sporophyte of the Fissidentales is similar to that of the Dicranales, and they may have some very ancient common ancestry, but the two groups would appear to have diverged ecologically with the onset of Paleozoic desert episodes.

Pottiales are often very seasonal mosses of exposed mineral soils or decaying rock.   Two adaptations to xeric conditions have proven successful—1) production of very short-lived, almost acaulescent, gametophyte on which a capsule containing long-lived spores is soon matured; and 2) a cytoplasmic adaptation to air- drying without loss of viability, perhaps by colloidal binding of sufficient water in the cytoplasm to prevent destruction of vital compounds, for periods of up to several years.   Some species in my laboratory from Guadalupe Island, Mexico, have sprouted from shoots placed in moist petri dishes after five years of storage as herbarium specimens.

Grimmiales are mostly blackish mosses growing on granitic or similar rocks in exposed sites such as dry tundra or in alpine regions.   As with the Pottiales, most fruit abundantly and some dispersal is accomplished by plant fragments. The developing pro- tonemata, as well as rhizoids, have been observed to etch the sur- face of highly polished granite and to thus become established on new surfaces.   Members of this family probably hold the longevity record for viability after dry herbarium storage of something over 20 years.

Orthotrichales are frequently epiphytic above the tree base and into the crown although some have adapted to near-neutral rock surfaces. Even though the cytoplasmic resistance to desiccation is somewhat less than for the two previous orders, it is nonetheless considerable.

Funariales are mostly ephemeral or annual mosses growing on exposed soil and reproducing quickly by abundant spore production.

Bryales are somewhat more persistent than Funariales although they normally produce abundant spores in season. Several genera are known to produce desiccation-resistant brood-bodies which both aid in dispersal and in carrying over fixed populations through progressive seasons.

Leucodontales and Hypnobryales are pleurocarpous orders which share eco-niches on an opportunistic basis in some temperate areas. In general, however, the Leucodontales are more abundant in tropical latitudes forming festoons of mosses in wet forests as opposed to Hypnobryales which seem much more at home on the forest floor. They form mats on decaying logs which provide a more constantly moist situation and a higher carbon dioxide concentration serving perhaps to enhance photosynthetic efficiency.

To summarize from a geobotanical point of view, in a broad sense, we are faced with a combination of facts and probabilities, which are at variance to some degree with some widely held points of view. New discoveries will surely modify, and hopefully reinforce, the key attributes of bryophyte origin and history. For elaboration of points not dealt with in this presentation, the reader is directed to two previous publications (Miller, 1974a,b). The origin and evolutionary history of the bryophytes may be summarized for the present as follows:

1) We must accept that the bryophytes are terrestrial vascular plants and have a long history which parallels that of the phylogenetically oldest tracheophytes on earth today.

2) The several major evolutionary lines lumped under Bryophyta—the Musci, Hepaticae and Anthocerotae—share only the common ancestry of the gene pool of the ancient rhyniophytes and zosterophyllophytes with bryophytes combining aspects of each of these lines.

3) Bryophytes were widely distributed and very diverse prior to the Permo-Triassic desert episodes which resulted in wholesale extinction of all but a few types which managed to adapt to arid situations.

4) With the rise of angiospermous vegetation this rag tag lot of bryophyte survivors was provided anew with a spectrum of mesic conditions and hospitable plant hosts followed by a second diversity explosion at lower taxonomic levels which paralleled otherwise that of the angiosperms.

5) Present day distributions and the systematic isolation of many bryophyte groups correlate well with the history of continental land masses and their climates as well as with parallel cases across the broad spectrum of other vascular plants.

Literature Cited

Banks, H. P. 1968a. The early history of land plants. pp.73-107. In Drake, E. T. (ed.) Evolution and Environment. Yale Univ. Press. New Haven.

_____. 1968b. The stratigraphic occurrence of early land plants and its bearing on their origin. pp. 721-730. In Oswald, D. H. (ed.) Proc. Internat. Symp. on the Devonian System. Calgary, Canada.

_____, and M. R. Davis. 1969. Crenaticaulis, a new genus of Devonian plants allied to Zosterophyllum, and its bearing on the classification of early land plants. Amer. J. Bot. 56: 436-449.

Bold, H. C. 1973. Morphology of Plants. 3rd edition. Harper and Row. New York.

Brown, J. T., and C. R. Robison. 1974. Diettertia montanensis, gen. et sp. nov., a fossil moss from the Lower Cretaceous Kootenai Formation of Montana. Bot. Gaz. 135: 170-173.

Finocchio, A. F. 1967. Pitting of cells in moss gametophores. Bull. Torrey Bot. Club 94: 18-20.

Gensel, P., A. Kasper and H. N. Andrews. 1969. Kaulangiophyton, a new genus of plants from the Devonian of Maine. Bull. Torrey Bot. Club 96: 261-276.

Hébant, C. 1964. Signification et évolution des tissus conducteurs chez les bryophytes. Nat. Monspeliensia Ser. Bot. 16: 79-86.

_____. 1967. Sur la comparison des tissus conducteurs des bryophytes et des plantes vasculaires. Compt. Rend. Acad. Sci. Paris Ser. D. 264: 901-903.

Heuber, F. M. 1971. Sawdonia ornata: a new name for Psilophyton princeps var. ornatum. Taxon 20: 641-642.

Kawai, I. 1971a. Systematic studies on the conducting tissue of the gametophyte in Musci (2). On the affinity regarding the inner structure of the stem in some species of Dicranaceae, Bartramiaceae, Entodontaceae and Fissidentaceae. Ann. Rept.

Bot. Gard. Fac. Sci. Univ. Kanazawa 4: 18-39.

_____. 1971b. Systematic studies on the conducting tissue of the gametophyte in Musci (3). On the affinity regarding the inner structure of the stem in some species of Thuidiaceae. Sci. Rept. Kanazawa Univ. 16: 21-60.

_____. 1971c. Systematic studies on the conducting tissue of the gametophyte in Musci (4). On the affinity regarding the inner structure of the stem in some species of Mniaceae. Sci. Rept. Kanazawa Univ. 16: 83-111.

Khanna, K. R. 1964. Differential evolutionary activity in bryo-phytes. Evolution 18: 652-670.

King, L. C. 1961. The palaeoclimatology of Gondwanaland during the Palaeozoic and Mesozoic Eras. pp. 307-331. In Nairn, A. E. M. (ed.) Descriptive Palaeoclimatology. Interscience. New York.

Lemoigne, Y. 1969. Observation d'archegones portes par des axes du type Rhynia gwynne-vaughanii Kidston et Lang. Existence de gametophytes vascularises au Devonien. Compt. Rend. Acad. Sci. Paris Ser. D. 267: 1655-1657.

_____. 1970. Nouvelles diagnoses du genre Rhynia et de l'espece Rhynia gwynne-vaughanii. Bull. Soc. Bot. France 117:307-320.

Merker, H. 1958. Zum fehlenden Gliede der Rhynienflora. Bot. Not. 111: 608-618.

Miller, H. A. 1974a. Rhyniophytina, alternation of generations, and the evolution of bryophytes. J. Hattori Bot. Lab. 38: 161-168.

_____. 1974b. Hepaticae through the ages. Revista Fac. Cien. Lisboa 17: 763-775.

Schopf, J. M. 1970. Relation of floras of the southern hemisphere to continental drift. Taxon 19: 657-674.

Schwarzbach, M. 1961. The climatic history of Europe and North America. pp. 255-291. In Nairn, A. E. M. (ed.) Descriptive Palaeoclimatology. Interscience. New York.

Steere, W. C. 1969. A new look at evolution and phylogeny in bryophytes. Current Topics in Plant Science 1969: 134-143.

MODERN AND PALEOCENE <u>METASEQUOIAS</u>:

A COMPARISON OF FOLIAGE MORPHOLOGY

Jerry L. Harr and Francis T. C. Ting

West Virginia University

Morgantown, West Virginia   26506

ABSTRACT

Silicified peat from the Paleocene Fort Union Group in North Dakota has been studied using thin-section and peel methods. The internal structure of foliage identified as <u>Metasequoia</u> <u>sp.</u> is well preserved and is compared to the internal morphology of <u>M. glyptostroboides</u> foliage.

A single vascular bundle runs the length of the leaf, with a resin canal located abaxial to it, similar to <u>M. glyptostroboides</u>. Two additional resin canals occur in the mesophyll just above the lower epidermis, while <u>M. glyptostroboides</u> forms resin canals in the leaf margins. The vascular bundle is enclosed in a sheath of parenchyma and fibers with an incomplete ring of large endodermal-like cells located outside the sheath. No transfusion tracheids have been seen in fossil material. In the fossil material, preservation of the mesophyll is insufficient to differentiate it into palisade and spongy tissues. Palisade mesophyll is dorsi-ventral in <u>M. glyptostroboides</u>. No stomata are found on the upper epidermis but are well preserved on the lower surface. The degree of distortion of plant tissues during peatification is readily apparent.

INTRODUCTION

The silicified peat utilized in this study is from the Williston Basin in western North Dakota. According to Laird (1956), this basin has been sinking slowly, and since the end of

109

the Pre-Cambrian approximately 500 million years ago,
approximately 15,000 feet of sediment have accumulated.

The Fort Union Group, of Paleocene age, which contains the
silicified peat, is composed of mostly non-marine sands, shales,
clays, and lignite deposited on a large alluvial plain (Laird,
1950) close to sea level. It is composed of four formations --
Tongue River, Sentinel Butte, Cannonball, and Ludlow (Royse,
1967). The majority of the silicified peat outcrops are along
the boundary between the Tongue River and the Sentinel Butte
Formations.

The Tongue River Formation was named from the Tongue River
in Wyoming and contains clinker beds where lignite was burned in
place. The Sentinel Butte Formation consists of usually darker-
colored sediments and has fewer lignite beds than the Tongue River
Formation (Laird, 1956). The Fort Union Group contains beds of
bentonitic clay which were formed from fine volcanic material that
probably blew in from the Rocky Mountains to the west, where vol-
canoes were erupting (Laird, 1956).

Silicifed peat is rare when compared with carbonate permineralized peat.

The famous Rhynie chert (Kidston and Lang, 1917-1921) from
the Devonian of Scotland is perhaps the best known example of a
peat permineralized by the activity of hot springs (Mackie,
1916), although the silicified peat described by Schopf (1970,
1971) from a Permian coal in Antarctica provides minute anatomical
details of the contained plants. A third occurrence has been
reported by Arnold (1963, 1964, 1970). He describes the anatomy
of several ferns preserved in an Eocene chert from the Clarno Formation in Oregon. Prior to his discovery near Medora, North
Dakota, Ting (1973) had studied the petrography and palynology of
a peat discovered in a silicified log from Center, North Dakota.

The material used in this study was initially discovered by
Francis Ting (1972).

A preliminary study of the palynology of the material was
published by Nichols and Ting in 1973. They reported a restricted
assemblage dominated by two genera, Taxodiaceaepollenites and
Laevigatosporites. Roland Brown (1962) sampled the same horizon
as the silicified peat in his paper "Paleocene Flora of the Rocky
Mountains and Great Plains." The descriptions of the plant fossils in Brown's paper were used to identify many of the plant
fossils collected at the silicified peat localities. Stanley
(1965) studied the Fort Union lignites in "Upper Cretaceous and
Paleocene Plant Microfossils from Northwestern South Dakota."

DESCRIPTIONS OF SILICIFIED PEAT LOCALITIES
WHERE METASEQUOIA WAS FOUND

During the summer of 1975, silicified peat or material
relating to it was collected at 17 localities from central to
western North Dakota (Figure 1). All of the localities except 7
and 14 were at or near the contact between the Tongue River and
Sentinel Butte Formations of the Fort Union Group.

Locality 5

Locality 5 is 6.5 kilometers north of New Salem in Morton
County. A silicified bed capped by a thin layer of peat contains
abundant Ginkgo and Metasequoia plus various angiosperms. The
shape of the silicified bed is that of a sand bar.

Locality 8

Locality 8 is located at a road cut east-southeast from
Medora (NW 1/4, Section 15, Township 139N, Range 101W). The

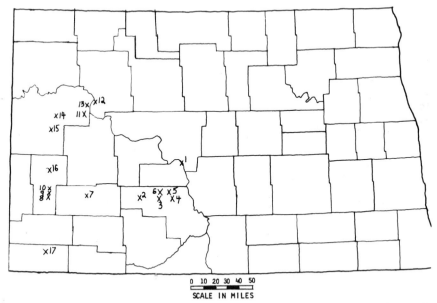

Figure 1: Map of North Dakota Showing Approximate Location of
Silicified Peat Outcrops.

silicified peat is intermittent, ranging from seven to
15 centimeters thick, with five centimeters of lignite above and
15 centimeters below. Two blocks of silicified peat have been
collected for thin sectioning. Plant fossils have also been collected from calcareous concretions which were abundant above the
silicified peat.

Locality 16

     Locality 16 is along Route 719 across from a benchmark
(Section 24, Township 143N, Range 101W) and is approximately
16 meters stratigraphically above the silicified peat at Medora.
Several large calcareous concretions are present, containing well
preserved plant fossils.

RECENT LEAVES AND TWIGS

     Commercial sections of modern conifer leaves and twigs were
purchased from the Triarch Biological Supply House. Sixteen
genera were examined and compared with fossil material from the
silicified peat. A list of conifer genera examined includes:
Abies, Cunninghamia, Ginkgo, Juniperus, Larix, Metasequoia, Picea,
Pinus, Podocarpus, Pseudotsuga, Sciadopitys, Sequoia, Taxodium,
Taxus, Thuja, and Tsuga. After the initial study of the silicified peat, all but one of the 16 genera were eliminated because
of anatomical differences with the fossil leaves in the silicified
peat. Examples of the foliage of Metasequoia glyptostroboides
were collected for serial sectioning and comparison with fossil
material.

Metasequoia glyptostroboides Hu and Chang

     Foliage (Figure 2) from Metasequoia glyptostroboides, the
only extant species of the genus Metasequoia, was collected from
the Missouri Botanical Gardens. The plant material was killed
and fixed in formalin-acetic-alcohol immediately upon collection.
Seven different wax blocks were prepared for sectioning. Serial
cross sections (Figure 3) and longitudinal sections of mature
leaves were cut along with serial cross sections of long and
short shoots with leaves attached.

     Mature leaves of Metasequoia have a longitudinal vascular
bundle with a median resin canal located abaxially (Sterling,
1949). Additional resin canals (Figure 3) occur at each margin
of the leaf, for a total of three. The vascular bundle belongs
to the collateral type and is unbranched. Two strands of

Figure 2:   Metasequoia glyptostroboides Hu and Chang.   Short
            Foliage Shoot Showing Shape and Opposite Arrangement
            of the Leaves.

transfusion tracheids appear lateral to the vascular bundle and
sometimes extend into the spongy mesophyll.

     The mature Metasequoia leaf has few cellular elements which
serve solely in a mechanical strengthening role. A thin strand
of collenchyma cells appears in the basal section of a mature
leaf (Figure 4), between the adaxial epidermis and the vascular
bundle, but is absent in sections near the apex of the leaf.
Collenchyma cells appear lateral to the median resin canal in
basal cross sections but are absent in sections near the apex of
the leaf. A distinct ring of endodermal cells is absent in the
extant Metasequoia, but a thin layer of parenchyma cells occurs
above and below the vascular bundle.   A single layer of palisade
mesophyll, located just under the adaxial epidermis, contains
numerous chloroplasts and is composed of densely packed elongate
cells.   Leaves having palisade mesophyll on the adaxial surface
only are described as dorsi-ventral (Figure 3). Cells comprising
the spongy mesophyll are irregular-shaped chlorenchyma cells containing chloroplasts fewer in number than the palisade cells.

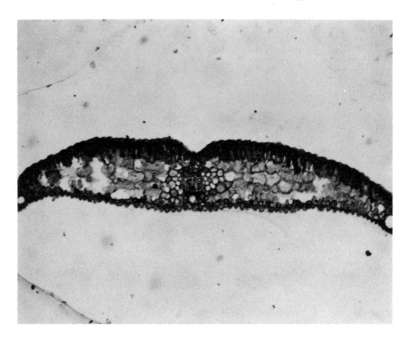

Figure 3:   Metasequoia glyptostroboides Hu and Chang.   Cross
            Section of a Mature Leaf Showing Three Resin
            Canals.   Slide 11-5.

     The three resin canals are schizogenous in origin, having a
well defined epithelium at the apex and midsection. But the epithelium is less well defined in the basal section of the leaves.
The cuticle of the epidermis is more prominent on the upper surface of the leaf. Metasequoia leaves are hypostomatic, having
stomata only on the lower epidermis.

     Metasequoia leaves are opposite and decussate on short and
long shoots (Morley, 1948). Each leaf twists at the node until
the leaf is parallel to the ground, giving the shoot a distichous
appearance (Figure 5). A cross section of the shoot stem consists
of a central pith surrounded by a layer of xylem and a layer of
phloem. When the xylem and phloem are interrupted by departing
leaf traces, the space is filled by cells that appear to be parenchyma. The cortex is composed mostly of parenchyma and is
photosynthetic.

     The median resin canal originates by the separation of the
border or epithelium cells and extends through the decurrent base
into the stem's cortex. The epidermis is cutinized and has

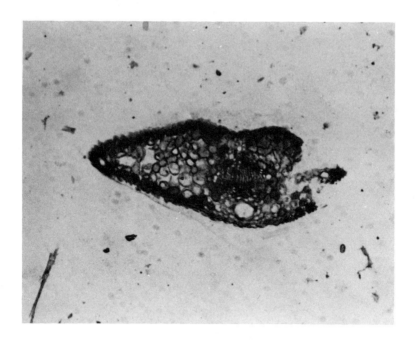

Figure 4:   Metasequoia glyptostroboides Hu and Chang.   Cross
            Section through the Base of a Mature Leaf.   Slide 12-1.

scattered collenchyma fibers just beneath.   Stomates occur on
decurrent bases of leaves but not on the epidermis of the stem.

Short shoots are borne in the axils of leaves on long
shoots.   Leaf bases on long shoots are constricted and appear
triangular or rounded in cross section.   They have a small vascular bundle and a large resin canal located between the vascular
bundle and the abaxial epidermis.

FOSSIL LEAVES AND TWIGS
FROM THE SILICIFIED PEAT

Plant fossils have been collected from silicified peat
localities in western North Dakota in the Paleocene Fort Union
Group.   Compression fossils of Metasequoia occidentalis are preserved in calcareous concretions and silicified beds.   With the
exception of Locality 16, all plant fossils were collected from
beds or concretions in contact with the silicified peat.

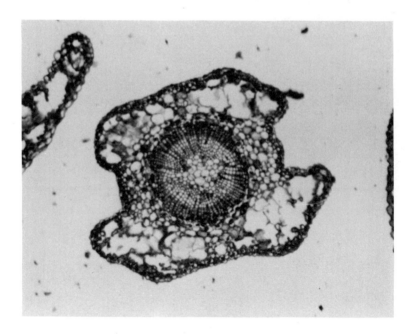

Figure 5:   Metasequoia glyptostroboides Hu and Chang.   Cross
            Section through a Short Shoot Showing Mode of
            Attachment of Leaves.   Slide 10-5c.

     The silicified peat was collected at 16 localities. The
peat samples were sectioned or peeled by the sheet method and
made into slides. The slides were then examined for plant structures. All of the fossil material described came from Locality 8.
Fossil foliage from this locality will be described and compared
to modern conifer leaves with similar structure.

     Slides of silicified peat from other localities contain
outlines of plant structures, but preservation is too poor to
identify individual plant tissues. Only leaves and twigs were
identified, with the other plant structures being left for later
study.

     Cross sections of various types of conifer leaves were
located on thin sections and peels of the silicified peat from
Locality 8 and grouped according to the tissue location and structure. Five major types of conifer leaves were discovered in the
silicified peat. Two of these groups were later subdivided into
four smaller groups based on leaf anatomy. Three of these five
groups were described and compared to modern Metasequoia.

Small twigs with leaves attached were also located on slides from the silicified peat and placed in two large groups based on opposite or non-opposite leaf arrangements. The Metasequoia type of leaf arrangement will be described and matched as closely as possible to its modern counterpart. An American Optical microscope (serial number 908708) now in the authors' possession was used to locate sections on slides used in this study.

Metasequoia occidentalis (Newberry) Chaney
(Chaney, 1951, pp. 171-263)
(Locality 5, Specimens 250507, 280511, 280521, 280524;
Locality 16, Specimens 281602, 281603, 281605,
281608, 281610, 281613, 281616)

Metasequoia occidentalis, the most abundant conifer in Tertiary rocks in western North America (Chaney, 1951), may or may not show branches with opposite foliage shoots due to partial shedding. Foliage branches usually have opposite and decussate foliage shoots, but short shoots are sometimes replaced by a large needle. Foliage shoots usually show slender stems and opposite and decussate leaves with decurrent bases (Chaney, 1951). Leaves are commonly rotated to form distichous or two-ranked shoots except at growing tips. Short shoots (Figure 6), the most common foliage unit, are deciduous and measure 1.0 by 3.5 centimeters.

Long shoots (Figure 7) have indeterminate growth, are usually not deciduous, and usually terminate branches of Metasequoia. Leaves on the long shoots are opposite and decussate and form sprays (2.5 by 6.0 centimeters).

M. occidentalis leaves are described as acicular, distichous, and monomorphic. They are rounded at the base, having a short petiole and a prominent midvein. The apex is mucronate to obtuse, and the leaf shape is lanceolate to elliptic. The base of a leaf has been so twisted at its point of attachment that contact is oblique, and the base is decurrent diagonally on the stem, often with an oblique trend. Leaves on long shoots measure 2.0 by 25.0 centimeters, while leaves on short shoots measure 1.0 by 6.0 centimeters.

Metasequoia occidentalis occurs at two of the silicified peat localities. At Locality 5, it occurs below the silicified peat, and at Locality 16, it occurs above but not immediately above the peat.

Figure 6: Metasequoia occidentalis (Newberry) Chaney. Short
           Foliage Shoot Bearing Opposite Leaves. Specimen 281603.

Foliage Type 1
(Medora Flora -- Silicified Peat)

Type Specimens

Slide Number	Coordinate-Ordinate Number
271030	43.5 x 131.4
271066	54.5 x 131.8
271016	38.0 x 136.8
271006	50.4 x 136.0

     Cross sections of foliage leaves assigned to Type 1 are
circular to elliptical and have a vascular bundle that occupies
30 percent of the cross section. The vascular bundle belongs to
the collateral type and has only one strand of phloem. A sheath
of thick-walled, fiber-like cells surrounds the vascular tissue.
The fiber-like cells are gradually replaced by thinner-walled
parenchyma-like cells as the area of the cross section increases.
An incomplete ring of large endodermal-like cells encloses the
vascular bundle.

Figure 7:   Metasequoia occidentalis (Newberry) Chaney.   Long
            Foliage Shoot Bearing Four Short Shoots.
            Specimen 281605.

        The mesophyll is poorly preserved and cannot be subdivided
into palisade and spongy layers. The median resin canal is absent
or not preserved in the Type 1 sections. Two marginal resin canals
are preserved in one section and located in the mesophyll just
above the abaxial epidermis. The epidermis is composed of two
layers, an outer layer bearing a cuticle and an inner or hypodermal
layer composed of thick-walled, fiber-like cells.

        Type 1 leaf sections increase in size and become more ellip-
tical in shape as the vascular bundle occupies a smaller percentage
of the area of the leaf cross sections. Changes in size, shape,
and structure of the leaf sections indicate that Type 1 sections
belong to a conifer leaf near the base or point of attachment of
the leaf.

        One section (Figure 8) from an acetate peel has well pre-
served xylem and phloem plus two marginal resin canals. The median
resin canal appears to be absent or not preserved in these sections.
The marginal resin canals (Figure 8) show a lining of intact cells

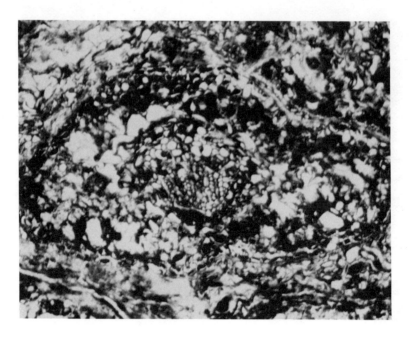

Figure 8:   Leaf Type 1 (<u>Metasequoia</u>). Best Preserved Section of
            Base of Fossil <u>Metasequoia</u> Leaf Showing Resin Canals.
            Slide 271066.

or epithelium, indicating a schizogenous origin for the resin
canals. Details for the xylem, phloem, and resin canals in the
composite drawing are taken from Figure 8. Cell structure of the
epidermis, hypodermis, and mesophyll for the composite drawing are
from another section. Since none of the sections of Type 1 has
all of the tissue preserved, a composite drawing was made
(Figure 9) to show a Type 1 leaf with all tissue preserved.

     Type 1 leaves increase in size and become flatter, while the
vascular bundle decreases in size and the fiber cells in the
bundle sheath are replaced by thin-walled cells. This would indicate that these sections are from farther up the leaf from the
point of attachment. Type 1 leaves intergrade with Type 4 leaves
and belong to the same generic group as Type 4.

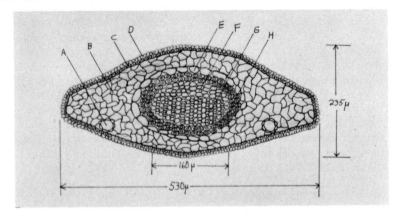

Figure 9: Composite Drawing Taken from Four <u>Metasequoia</u>
           (Type 1) Leaves near the Base or Point of Attachment
           of the Leaf.

A - Resin Canal   E - Endodermis
B - Mesophyll   F - Xylem
C - Hypodermis   G - Phloem
D - Cuticle   H - Fibers

Foliage Type 2
(Medora Flora -- Silicified Peat)

Type Specimens

Slide Number	Coordinate-Ordinate Number
271010	34.5 x 137.5
271013	39.0 x 129.3
271005	54.5 x 137.2
271030	43.5 x 131.4

Cross sections of conifer leaves that are triangular in
outline are described as Type 2 leaves. This group is later subdivided into two smaller groups, leaves that are triangular

because of decomposition or angle of section (Type 2A) and leaves
that are triangular because of folding or deformation (Type 2B).

Type 2A sections have a vascular bundle like that of Type 1
and a similar mesophyll. The epidermis is incomplete, so the true
shape of the leaves is unknown. The section's present shape is
triangular due to decomposition or abrasion of the mesophyll during
deposition. The bundle sheath is composed of thin-walled cells
with a few thick-walled fiber cells surrounded by large endodermal-
like cells, the same type of vascular bundle that is transitional
between Type 1 and Type 4 leaves.

Some Type 2A sections are oblique and have a triangular outline. The xylem cells are cut at an angle and show annular secondary thickening. A wedge of dark-filled xylem tracheids appears
to divide the xylem strand in half, a characteristic found in Type 4
foliage. The epidermis is intact around the section, which if
reorientated parallel to the leaf axis would have the elliptical
outline of a Type 4 leaf.

Several triangular leaf sections (Figures 10 and 11), classified as Type 2B, represent distorted or folded sections with

Figure 10: Leaf Type 2. Oblique Section of a Leaf with Epidermis
            Folded into the Mesophyll. Slide 271005.

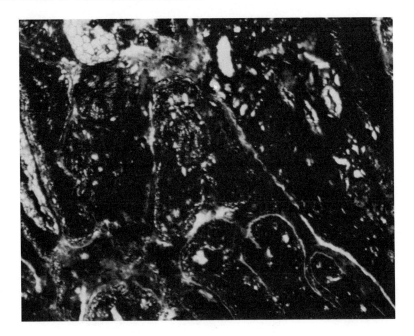

Figure 11:   Leaf Type 2 (<u>Metasequoia</u>).   Cross Section of a Leaf
             Having Epidermis Folded into Mesophyll.   Slide 271030.

considerable disturbance of cell structure in the mesophyll.   Upon
examination of the xylem of one distorted section (Figure 10),
annular secondary thickening of the tracheids is clearly visible,
indicating an oblique as well as a folded section.   Tracing the
epidermis of the leaf section (Figure 10) shows a large section
folded and buckled into the mesophyll on the shortest side of the
triangular section.   A considerable amount of cellular disturbance
in the mesophyll has also occurred during the folding of the leaf.

        Another leaf section (Figure 11) is folded in the same manner
as the previous section (Figure 10) but is perpendicular to the
axis of the leaf.   In the xylem, a wedge of darkened tracheids
appears to divide the xylem strand in half, similar to Figure 10
and Type 4 leaves.   The epidermis is complete, with a large
in-folded section and considerable cellular disruption on the opposite side of the vascular bundle.   All of the fiber cells in the
bundle sheath have been replaced by thin-walled parenchyma-like
cells, a characteristic of Type 4 leaves.   A small resin canal
with four epithelial cells is located in the mesophyll just above
the lower epidermis, further illustrating the relationship between
Type 2B and Type 4 leaves.

Foliage Type 4
(Medora Flora -- Silicified Peat)

Type Specimens

Slide Number	Coordinate-Ordinate Number
271006	50.8 x 134.8
271029	46.0 x 138.5
271029	47.7 x 147.1
271043	37.6 x 131.0
271063	53.0 x 145.5
271067	30.3 x 144.3

Foliage assigned to Type 4 shows elliptical cross sections bearing three resin canals. Type 1 and Type 4 leaves intergrade. The criterion for separating the two types is the presence of fiber-like cells in the vascular bundle of Type 1 and their absence in Type 4. Because of intergrading structure, Type 1 and Type 4 sections are from the same type of leaf. Type 1 sections are from near the base or point of attachment, while Type 4 represents sections from the midpoint of the leaf. Type 1 and Type 4 leaves are the most common leaf sections in the silicified peat.

Like Type 1 leaves, Type 4 occurs in different stages of preservation. Sometimes different parts of the leaf are preserved while others are not. For this reason, the description of Type 4 is based on six different sections, and a composite drawing (Figure 12) has been prepared. Of all of the commercially prepared sections which have been studied and compared with Type 4, only two genera, Metasequoia and Sequoia, have similar internal structure. Type 4 leaves will be described and compared with the previous description of Metasequoia.

Type 4 leaf sections are elliptical (Figure 13) in outline unless visibly distorted. Measurements are taken from average Type 4 sections. Variations in width of leaves are from 830 to 1,200 microns. Leaf thickness varies from 130 to 160 microns. The vascular bundle, not including the bundle sheath, ranges from 110 to 130 microns.

Type 4 leaves (Figure 14) have a collateral vascular bundle with a curved cambium. The size of the vascular bundle in relation to the thickness of the leaf is similar in Metasequoia and Type 4 leaves. The thickness of the vascular bundle in Sequoia occupies about one-half the thickness of the leaf, and the remainder of the space is filled with mesophyll. However, in Metasequoia and Type 4 leaves, almost the entire thickness of the leaf is occupied by the vascular bundle, leaving very little space for mesophyll.

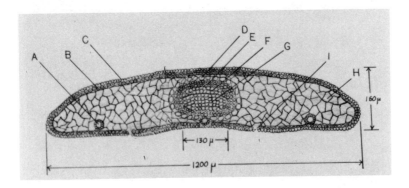

Figure 12:   Composite Drawing of Six Sections of Metasequoia
              Leaves (Type 4) near the Midsection of a Mature Leaf.

         A - Resin Canal   F - Xylem
         B - Hypodermis   G - Phloem
         C - Cuticle   H - Mesophyll
         D - Endodermis   I - Stomate
         E - Xylary Parenchyma

       In a section with well preserved vascular tissue, xylem is
very easy to differentiate from phloem. Xylem cells occur adaxial
to the phloem and are thicker walled and larger, ranging from
11 to 22 microns. Phloem cells are smaller than xylem, measuring
from six to 11 microns in diameter.

       A sheath of parenchyma cells surrounds the vascular bundle
instead of the fiber cells of Type 1 leaves. The absence of these
fiber cells separates Type 1 from Type 4 foliage. Surrounding the
bundle sheath is a ring of large endodermal-like cells which in
some specimens is incomplete. In Type 4 leaves, collenchyma cells
in the bundle sheath are gradually replaced at random by parenchyma
cells, but in modern Sequoia and Metasequoia, collenchyma cells are
replaced by parenchyma cells from the lateral sides of the bundle
sheath.

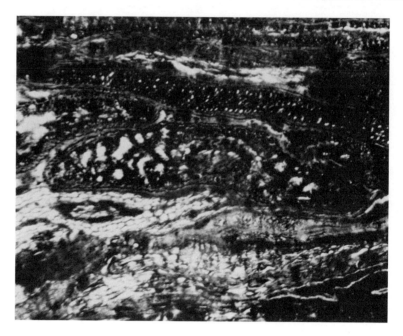

Figure 13: Leaf Type 4 (Metasequoia). Cross Section of a Well
            Preserved Leaf with Marginal and Median Resin Canals.
            Slide 271029.

Type 4 leaves contain three resin canals of the schizogenous type (Figure 13). The median resin canal is located between the vascular bundle and the abaxial epidermis. Two marginal resin canals are located in the mesophyll above the abaxial epidermis, similar to modern Sequoia. The mesophyll cannot be subdivided into palisade and spongy in any of the leaf sections.

A single layer of hypodermal cells lies just under the epidermis (Figure 13) but consists of a double layer between the vascular bundle and the adaxial epidermis. Modern Metasequoia does not contain a hypodermis or multi-layered epidermis, but Sequoia has a hypodermis, composed of thick-walled cells under the adaxial epidermis only. Type 4 leaves contain stomata on the abaxial epidermis only, similar to Metasequoia and Sequoia.

Type 4 leaf sections have some characteristics in common with both modern Sequoia and modern Metasequoia, but Type 4 leaves are assigned to the genus Metasequoia because Type 4 leaves are found with an opposite mode of attachment (Figure 15) to twigs and the large size of the vascular bundle, both distinguishing characteristics of Metasequoia.

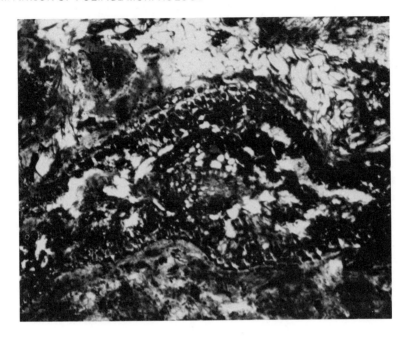

Figure 14: Leaf Type 4 (Metasequoia). A Distorted Cross Section
            of a Metasequoia Leaf with Well Preserved Vascular
            Bundle. Slide 271006.

        Since only one species of Metasequoia (M. occidentalis) and
no Sequoia species have been found at any of the silicified peat
localities, Type 1 and Type 4 leaves probably represent sections of
leaves from Metasequoia occidentalis, which have been described
from compressed foliage. Absolute proof of this connection will
have to wait for better preserved compressions to be studied.

                            Twig Type 1
                (Medora Flora -- Silicified Peat)

                          Type Specimens

        Slide Number   Coordinate-Ordinate Number
          271005   43.4 x 136.3
          271043   61.8 x 129.8

        Sections of small twigs with opposite-arranged leaves which
have Metasequoia leaves attached have been classified as Type 1.
One twig section (Figure 15) has oppositely attached leaves with

Figure 15:   Twig Type 1.   Twig (Metasequoia) Showing Oppositely
             Attached Leaves with Decurrent Bases.   Slide 271043.

opposite decurrent bases almost identical to modern Metasequoia
(Figure 5).   The arrangement of fossil Metasequoia leaves around a
twig and the pattern of xylem and pith rays in the twig can be
seen.   In fossil Metasequoia, leaves lie on opposite sides of the
twig, compressed in against the twig, slightly overlapping to show
the decussate arrangement.

                         SUMMARY AND CONCLUSIONS

      Fossil leaves and twigs belonging to Metasequoia that have
been discovered in the silicified peat have been described and compared with modern Metasequoia.   Type 1 leaves intergrade with Type 4.
The fiber-like cells surrounding the vascular tissue in Type 1
leaves are gradually replaced by the parenchyma tissue in Type 4
leaves.   Type 1 and Type 4 leaves are assigned to the genus
Metasequoia based on the structure of the vascular bundle, its size
in relation to the leaf cross section, and the opposite mode of
attachment of the leaves.

Type 4 leaves also have several characteristics of Sequoia, particularly the size and position of the three resin canals and a multi-layered epidermis or hypodermis.

Type 2 leaves contain some abraded leaves which could not be identified and some triangular leaves that are identified as folded Type 4 leaves.

Type 1 twigs with Metasequoia leaves in an opposite mode of attachment are rare in the peat. Metasequoia leaves are common in the peat but not attached to twigs. Since Metasequoia has deciduous twigs, the absence of very many twigs of Metasequoia with leaves attached is a good indication that Metasequoia could have lived near the edge of the swamp or just upland and may have been transported just a short distance into the swamp.

ACKNOWLEDGMENTS

The authors wish to sincerely thank Dr. Herald D. Bennett and Mr. William H. Gillespie, who offered valuable assistance and suggestions relating to this paper. We would also like to thank the Missouri Botanical Gardens and the Theodore Roosevelt National Memorial Park for permission to collect samples. Field expenses for the summer of 1975 were supported by NSF Grant DES 74-24518. We also wish to thank Bays Mountain Park, Kingsport, Tennessee, which has kindly agreed to act as a repository for the fossils used in this study.

REFERENCES

Arnold, C. A. 1970. Eocene Age Plant Bearing Chert in Northwestern United States and Canada. GSA Abstracts with Programs, Vol. 2, No. 6, pp. 373-374.

Arnold, C. A., and Daugherty, Lyman H. 1963. The Fern Genus Acrostichum in the Eocene Clarno Formation of Oregon. Contributions to the Museum of Paleontology, University of Michigan, Vol. 18, No. 13, pp. 205-227.

Arnold, C. A., and Daugherty, Lyman H. 1964. A Fossil Dennstaedtioid Fern from the Eocene Clarno Formation of Oregon. Contributions to the Museum of Paleontology, University of Michigan, Vol. 19, pp. 65-88.

Brown, R. W. 1962. Paleocene Flora of the Rocky Mountains and Great Plains. USGS Professional Paper No. 375.

Chaney, R. W. 1951. A Revision of Fossil Sequoia and Taxodium in
    Western North America Based on the Recent Discovery of
    Metasequoia. American Philosophical Society Transactions,
    Vol. 40, Part 3, pp. 171-263.

Kidston, Robert, and Lang, W. H. 1917-1921. On Old Red Sandstone
    Plants Showing Structure, from the Rhynie Chert Bed. Royal
    Society of Edinburgh Transactions, Vol. 51, pp. 761-784;
    Vol. 52, pp. 603-627, 643-680, 831-854, and 855-902.

Laird, Wilson M. 1950. Geology of the South Unit, Theodore
    Roosevelt National Memorial Park. North Dakota Geological
    Survey Bulletin No. 25.

Laird, Wilson M. 1956. Geology of the North Unit, Theodore
    Roosevelt National Memorial Park. North Dakota Geological
    Survey Bulletin No. 32.

Mackie, William. 1916. The Rock Series of Craigbeg and Ord Hill,
    Rhynie. Royal Society of Edinburgh Transactions, Vol. 50,
    pp. 205-236.

Morley, Thomas. 1948. On Leaf Arrangement in Metasequoia
    glyptostroboides. Proceedings of the NAS, Vol. 34,
    pp. 574-578.

Nichols, D. J., and Ting, F.T.C. 1973. Palynology of Paleocene
    Petrified Peat (abstract). American Association of Strati-
    fied Palynologists Annual Meeting.

Royse, C. F., Jr. 1967. Tongue River-Sentinel Butte Contact,
    Western North Dakota. North Dakota Geological Survey
    Report No. 45.

Schopf, J. M. 1970. Petrified Peat from a Permian Coal Bed in
    Antarctica. Science, Vol. 169, pp. 274-277.

Schopf, J. M. 1971. Notes on Plant Tissue Preservation and
    Mineralization in a Permian Deposit of Peat from Antarctica.
    American Journal of Science, Vol. 271, pp. 522-543.

Stanley, E. A. 1965. Upper Cretaceous and Paleocene Plant Micro-
    fossils from Northwestern South Dakota. Bulletin of
    American Paleontology, Vol. 49, No. 22, pp. 179-384.

Sterling, Clarence. 1949. Some Features in the Morphology of
    Metasequoia. American Journal of Botany, Vol. 36,
    pp. 461-470.

Ting, F.T.C.   1972.   Petrified Peat from a Paleocene Lignite in
     North Dakota.   Science, Vol. 177, pp. 165-166.

Ting, F.T.C.   1973.   Petrology and Palynology of a Silicified
     Sapropelic Peat from a Paleocene Lignite Bed in North
     Dakota.   Geoscience and Man, Vol. 7, pp. 65-66.

A MIDDLE PENNSYLVANIAN NODULE FLORA FROM CARTERVILLE, ILLINOIS

Robert A. Gastaldo

Southern Illinois University

Carbondale, Illinois 62901

ABSTRACT

As early as 1875 the occurrence of small, irregularly-shaped
nodules of pyritiferous clay in which fossil plants could be found,
had been reported in the roof shales of the Herrin (No. 6) Coal at
Carterville, Illinois. Active collection of these plant bearing
nodules began nearly 15 years ago, and at present over 2500 specimens
are curated in the Southern Illinois University paleobotanical
collection. The impression-compression ironstone concretion flora
is Middle Pennsylvanian in age, assigned to the Carbondale Formation,
Kewanee Group. The completed floristic survey encompasses 1,475
specimens delegated to 24 genera and 52 species. Specimens not able
to be placed in a taxonomic rank were assigned to form status. The
flora is dominated (44% of all specimens assignable to a generic
status) by Filicalean and Marattialean elements of the genus
Pecopteris Brongniart. Medullosan pteridosperm taxa, *Neuropteris*
(Brongniart) Sternberg, *Alethopteris* Sternberg, *Odontopteris*
Sternberg, and *Callipteridium* Weiss compose nearly 25% of the flora.
Few Lyginopterid elements have been encountered. Calamitean
components are relatively abundant in the forms of *Annularia*
Sternberg, *Asterophyllites* Brongniart, and *Calamites* Suckow, while
Sphenophyllalean taxa are rare. Isolated sporophylls dominate
the Lepidodendralean aspect of the flora. The assemblage exhibits
an abundance of Upper Allegheny and Lower Conemaugh plants, and
has been equated to the Upper Kittanning Coal of the Appalachian
Region.

INTRODUCTION

Modern intensified study of the American Pennsylvanian megafossil flora is needed. Present concepts on American floral assemblages have been derived from the monumental works of Lesquereux (1879, 1880, 1884), Fontaine and I. C. White (1880), D. White (1899), and others. Over the past 40 years several manuals for the amateur (Janssen, 1939; Langford, 1958, 1963; Gillespie, Latimer, and Clendenning, 1966), and monographic works for the professional appear (ie. Janssen, 1940; Abbott, 1958; Read and Mamay, 1960; Cridland, Morris, and Baxter, 1963; Basson, 1968; Darrah, 1969). Methods and concepts which are applied to the study of American Carboniferous compression-impression floras have originated, and are still being originated with European paleobotanists (ie. Bertrand, 1930, 1932; Corsin, 1932, 1951; Crookall, 1955, 1959, 1964, 1966; Danze, 1956; Boersma, 1972). With the "completion" of the major task of floral identification, American paleobotanists concerned with the Pennsylvanian began turning their attention to other floristic problems. Since 1920, the major thrust of Pennsylvanian research has been directed toward the anatomy and morphology of floral components. Characterization of floral assemblages in the American Carboniferous are derived from intensive and complete studies. There is a continuing need for examination of undescribed floras, as well as reexamination of older described floras, in order to best assess plant assemblages, populations, diversity and range of the coal swamp taxa.

NOTES ON FLORISTIC ELEMENTS

The present study describes a megafossil flora from the Middle Pennsylvanian of Williamson County, southern Illinois. The Illinois State Geological Survey map of Shipping Coal Mines in Illinois (Cady, 1947) denotes the strip mine, S-13, from which all the plant material is collected, as located west of the Carterville township. The coordinates for the mine on the SE 4/ Herrin 15' Quadrangle are 4 - T9S - R1E. Hibbert and Eggert (1965) report the coal at the locality as the Herrin (No. 6) Coal Member which is placed within the Carbondale Formation of the Kewanee Group (Kosanke, Simon, Wanless, and Willman, 1960), (Fig. 1). The fossil material is collected from the overlying strata which had been removed during active mining. The plant fossils are found within "ironstone concretions" similar in character to those of the Francis Creek Shale Flora (Mazon Creek). Schopf (1975) describes this mode of preservation as authigenic cementation. Collection of the fossil material has been sporadic over the past fifteen years, and at present there are over 2500 specimens curated in the SIU paleobotany collection. The present study encompasses 1,475 specimens, 1,100 of which are assignable to a generic status, and 599 to a specific rank (Fig. 2).

```
┌─────────────────────────────────────────────────┐
│              McLEANSBORO  GROUP                   │
├──────┬──────────┬───────────────────────────────┤
│      │          │ Danville (No. 7) Coal          │
│      │          │ Galum LS.                      │
│      │          │ Allenby Coal                   │
│      │    FM.   │ Bankston Fork LS.              │
│      │          │ Anvil Rock SS.                 │
│      │          │ Conant LS.                     │
│      │ DALE     │ Jamestown Coal                 │
│      │          │ Brereton LS.                   │
│GROUP │ CARBON   │ Herrin (No. 6) Coal            │
│      │          │ Vermilionville SS.             │
│KEWANEE│         │ Briar Hill (No. 5A) Coal       │
│      │          │ St. David LS.                  │
│      │          │ Harrisburg (No. 5) Coal        │
│      │          │ Hanover LS.                    │
│      │          │ Summum (No. 4) Coal            │
│      │          │ Roodhouse Coal                 │
│      │          │ Shawneetown Coal               │
│      │          │ Colchester (No. 2) Coal        │
│      ├──────────┴───────────────────────────────┤
│      │            SPOON  FM.                      │
├──────┴──────────────────────────────────────────┤
│              McCORMICK  GROUP                      │
└──────────────────────────────────────────────────┘
```

Figure 1. Classification of Pennsylvanian Stratigraphy of the Kewanee Group in southern Illinois (after Kosanke, Simon, Wanless, and Willman, 1960).

Lycopsida
(Figures 5-21)

The largest number of described taxa occur within the Lepidodendrales which are represented by fragmentary specimens composing 16% of the flora (based on 1,100 specimens). *Lepidodendron lanceolatum* Lesquereux, *Lepidodendron latifolium* Lesquereux, *Lepidodendron clypeatum* Lesquereux, and *Lepidodendron scutatum* Lesquereux have been identified from the Carterville locality. Lesquereux (1880) states that *Lepidodendron scutatum* could possibly be a variation of *Lepidodendron lanceolatum* except for the presence of short and narrow, linear leaves. Wagner (1962) remarks that *Lepidodendron scutatum* is characteristic of the Stephanian A of Spain, and is not found below that stratigraphic unit. The great abundance of taxa is due to the large number of disarticulated sporophylls including *Lepidostrobopsis* cf. *missouriensis* (White) Abbott, *Lepidostrobophyllum lanceolatum* (Lesquereux) Chaloner, *Lepidostrobophyllum brevifolium* (Lesquereux) Chaloner, *Lepidostrobophyllum hastatum* (Lesquereux) Chaloner, *Lepidostrobophyllum fallax* (Lesquereux) comb. nov., and *Lepidostrobophyllum* cf. *tumidum* (Lesquereux) comb. nov. *Lepidostrobophyllum fallax* (Lesquereux) comb. nov. and *Lepidostrobophyllum* cf. *tumidum* (Lesquereux) comb. nov. agree with the delimitations proposed by Lesquereux and are assignable to *Lepidostrobophyllum* Hirmer, rather than another sporophyll taxon. *Lepidocarpon major* (Brongniart) Hemingway is the most frequently encountered form of disarticulated sporophyll, and accounts for almost 5% of the flora.

Other Lepidodendralean elements encountered include a fragmentary cone, *Lepidostrobus ornatus* Brongniart, fragmentary *Lepidophloios laricinus* Sternberg, *Lepidophylloides* spp. Snigirevakaya, *Lepidocystis* spp. Lesquereux, *Lycopodites* spp. Lindley and Hutton, and *Stigmaria ficoides* Sternberg. Darrah (1969) remarks that within the form genus *Lycopodites*, some species may truely represent a creeping rhizomatous herbaceous plant similar in general habit to the extant *Lycopodium*, may be referable to small twigs of *Lepidodendron*, or may be juvenile sporophytes. The presence of this form genus is noted, but the specimens from Carterville shed little light on this problem.

GENERA based on 1100 specimens	SPECIES based on 599 specimens
Pecopteris 44%	Pecopteris lamuriana 14.7%
Neuropteris 20%	Neuropteris scheuchzeri f. scheuchzerii comb. nov. 7%

Lepidocarpon,
Lepidostrobophyllum,
Lepidostrobopsis 10%
	Neuropteris ovata complex 6.6%
	  (N. ovata f. typica ,
	   N. ovata f. flexuosa ,
	   N. ovata f. vermicularis)

Annularia, Asterophyllites, Sphenophyllum 9.5%	Pecopteris unita 6%

Pteridosperm foliage 4.6%
  (Alethopteris,
   Callipteridium,
   Odontopteris,
   Renaultia)
	Neuropteris scheuchzeri
	    f. decipiens comb. nov. 5%

	Lepidocarpon major 4.8%
Calamites 4.4%	Pecopteris cistii 4%
Lepidodendron, Lepidophylloios 2.9%	Annularia stellata 4%
	Neuropteris rarinervis 3%
	Lepidostrobophyllum lanceolatum 2%
	Annularia radiata 2%
	Asterophyllites equisetiformis 2%
	Calamites suckowii 2%
	Renaultia (Sphenopteris) chaerophylloides 1%
	Alethopteris serlii 1%

Figure 2. Generic and specific percent composition of the floristic assemblage at the Carterville, Illinois locality.

The Sigillariaceae is represented by *Asolanus campotaenia* Wood. This taxon has been assigned to this family by Chaloner (1967) and Darrah (1969).

Sphenopsida
(Figures 22-28; 31)

The Sphenopsida account for 14% of the flora assignable to Class rank (N = 1,100). The Calamitales are the prevailing Sphenopsid element at the site and are present in the forms of *Calamites carinatus* Sternberg, *Calamites suckowi* Brongniart, *Calamites cistii* Brongniart, and *Calamites undulatus* Sternberg. Although pith casts of these taxa are fairly abundant, foliage dominates. *Annularia radiata* (Brongniart) Sternberg, *Annularia stellata* (Schlotheim) Wood, and *Asterophyllites equisetiformis* Schlotheim account for over 8% of those taxa assignable to a species rank. A calamitalean cone, *Paracalamostachys cartervilii* Hibbert and Eggert, was first described from the Carterville site, and to date, one additional specimen has been collected.

The Sphenophyllales are very rare and fragmentary. Eleven specimens have been collected of which two have been assigned to *Sphenophyllum* cf. *emarginatum* Brongniart. It appears as if these plants were not in the immediate vicinity of the basin of deposition and may have been transported.

Leptosporangiopsida
(Figure 30)

Pecopteris (Senftenbergia) plumosa Artis (*Pecopteris dentata* Brongniart) has been placed in the Schizaceae due to the presence of an annulus on the sporangium (Corsin, 1951). This taxon is often encountered at the Carterville locality and composes almost 2% of those specimens assignable to a species status. Recently, Jennings and Eggert (1972) and Stidd (1974) have remarked that the *Senftenbergia*-type fructification may belong to an *Ankryopteris*-type plant.

Eusporangiopsida
(Figures 32-41)

Ten taxa, subdivided into four groups (Figure 3) are Marattialean representatives and compose the largest percentage of the flora. *Pecopteris (Asterotheca) lamuriana* Heer is the most frequently collected form and represents 14.7% of those specimens assignable to a species rank. Although this taxon is infrequently mentioned in North American literature, Darrah (1969) states that older collections possess many specimens of *Pecopteris lamuriana* Heer

Group <u>Pecopteris</u> <u>plumosa</u>-<u>dentata</u> Corsin
  <u>P</u>. <u>plumosa</u> Artis (<u>P</u>. <u>dentata</u> Brongniart)

Group <u>Pecopteris</u> <u>bucklandi</u> Darrah
  <u>P</u>. <u>bucklandi</u> Brongniart
  <u>P</u>. <u>cisti</u> Brongniart
  <u>P</u>. <u>daubreii</u> Zeiller

Group <u>Pecopteris</u> <u>arborescens</u>- <u>obliqua</u> Corsin
  <u>P</u>. <u>arborescens</u> (Schlotheim) Brongniart
  <u>P</u>. cf. <u>cyathea</u> Schlotheim

Group <u>Pecopteris</u> <u>miltoni</u> Corsin
  <u>P</u>. cf. <u>miltoni</u> Artis
  <u>P</u>. <u>lamuriana</u> Heer
  <u>P</u>. <u>oreopteridia</u> Schlotheim
  <u>P</u>. <u>psuedovestita</u> D. White

Group <u>Pecopteris</u> <u>unita</u> Corsin
  <u>P</u>. <u>unita</u> Brongniart

Figure 3. The assemblage of *Pecopteris* Brongniart at the
Carterville locality. Leptosporangiopsida are represented by Group
Pecopteris plumosa-dentata Corsin, while Eusporangiopsida are
represented by Group *Pecopteris bucklandi* Darrah, Group *Pecopteris
arborescens-obliqua* Corsin, Group *Pecopteris miltoni* Corsin, and
Group *Pecopteris unita* Corsin.

which have been identified as *Pecopteris villosa* Brongniart,
Pecopteris abbreviata Brongniart, *Pecopteris vestita* Lesquereux,
and *Pecopteris miltoni* Artis. It is noteworthy to recognize the
increased diversity of *Pecopteris* Brongniart at the Carterville
site with the presence of cyatheoid forms of Group *Pecopteris
arborescens-obliqua* Corsin, as well as the Group *Pecopteris
bucklandi* Darrah. *Pecopteris* (*Ptychocarpus*) *unita* Brongniart is
second in abundance of the pecopterid forms and represents 6% of
those specimens assignable to a species rank.

Cycadopsida
(Figures 42-59)

Eighteen taxa are assignable to the Cycadopsida and compose
25% of the flora (N = 1,100). Although Lyginopterid elements
are encountered, Medullosan components prevail. The most abundant
Medullosan genus is *Neuropteris* (Brongniart) Sternberg (Figure 4).
Nine taxa are recognized in the Carterville assemblage, of which
two dominate, *Neuropteris scheuchzeri* Hoffman forma *scheuchzeri*
comb. nov., and *Neuropteris scheuchzeri* Hoffman forma *decipiens*
(Lesquereux) comb. nov. These two taxa account for 12% of those
specimens assignable to a species status. There has been much
controversy associated with the placement of *Neuropteris decipiens*

Group <u>Neuropteris</u> <u>heterophylla-tenuifolia</u> Crookall

<u>N</u>. heterophylla Brongniart
<u>N</u>. <u>rarinervis</u> Bunbury
<u>N</u>. cf. <u>missouriensis</u> Lesquereux

Group <u>Neuropteris</u> <u>ovata-obliqua</u> Crookall

<u>N</u>. <u>ovata</u> Hoffman forma <u>typica</u> Crookall
<u>N</u>. <u>ovata</u> Hoffman f. <u>flexuosa</u> (Sternberg) Crookall
<u>N</u>. <u>ovata</u> Hoffman f. <u>vermicularis</u> (Lesquereux) Darrah

Distinctive Species

<u>N</u>. <u>scheuchzeri</u>(Hoffman)
   forma <u>scheuchzeri</u> comb. nov.
<u>N</u>. <u>scheuchzeri</u> (Hoffman)
   forma <u>decipiens</u> (Lesquereux) comb. nov.
<u>N</u>. <u>anomala</u> Lesquereux

Figure 4. The assemblage of *Neuropteris* (Brongniart) Sternberg
at the Carterville locality.

Lesquereux. Lesquereux (1880) separated *Neuropteris decipiens* from
Neuropteris hirsuta (= *Neuropteris scheuchzeri*) on the basis of
the character of the lateral veins. Crookall (1959) places
Neuropteris decipiens in synonymy with *Neuropteris scheuchzeri* due
to their similarities in character, which have been noted by a
variety of authors, including habit, pinnule form, nervation, and
villosity. Crookall also implies that *Neuropteris decipiens* is a
varietal type of *Neuropteris scheuchzeri*, a species notorious for
its polymorphism. Darrah (1969) remarks that the relationship
of *Neuropteris decipiens* to *Neuropteris scheuchzeri* is similar to
that of *Neuropteris vermicularis* to *Neuropteris ovata*, but retains
Neuropteris decipiens as a distinct species. The retention of
Neuropteris decipiens is based on characters which include: the
great size of pinnules which seldom show an acuminate tip; venation
which is fine, yet distinct (27–29 veins per cm. of margin length,
as opposed to a consistent number of 38–44 veins per cm. of margin
length in *Neuropteris scheuchzeri*); Odontopterid pinnules usually
intercalary of extreme variable form not found in association
with *Neuropteris scheuchzeri*; the presence of blunt, short-lobed,
or subcrenulate pinnules; and a highly restricted distribution
(Carbondale Formation of Illinois).

The specimens assignable to *Neuropteris decipiens* Lesquereux
as interpreted by Darrah (1969) have been placed into a new
combination because it is believed that the relationship of the
taxon to *Neuropteris scheuchzeri* Hoffman is strong due to their
common characters. These characters include: the presence of
trifoliate pinnae; the presence of stiff hairs upon the lower surface;
the midvein of the large pinnules ascending nearly to the apex; and
curved lateral veins which dichotomize four (occasionally five)
times. It is also believed that *Neuropteris decipiens* can not be

Figure 5. *Lepidodendron lanceolatum* Lesquereux. 663.185.
Figure 6. *Lepidodendron latifolium* Lesquereux. 452.
Figure 7. *Lepidodendron clypeatum* Lesquereux. JJ 1.47.
Figure 8. *Lepidodendron scutatum* Lesquereux. 534.
Figure 9. *Lepidophylloides laricinus* Sternberg. 425.

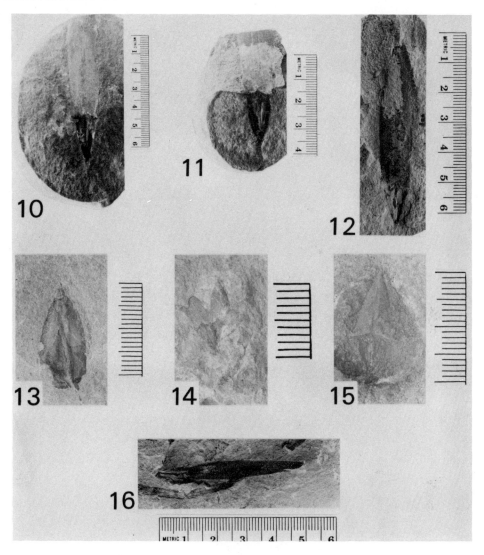

Figure 10. *Lepidocarpon major* (Brongniart) Hemingway. JJ 2.83.

Figure 11. *Lepidostrobopsis* cf. *missouriensis* (White) Abbott. JJ 14.46.

Figure 12. *Lepidostrobophyllum fallax* (Lesquereux) comb. nov. 484.

Figure 13. *Lepidostrobophyllum* cf. *tumidum* (Lesquereux) comb. nov. JJ 7.82.

Figure 14. *Lepidostrobophyllum brevifolium* (Lesquereux) Chaloner. JJ 2.213.

Figure 15. *Lepidostrobophyllum hastatum* (Lesquereux) Chaloner. JJ 8.40.

Figure 16. *Lepidostrobophyllum lanceolatum* (Lesquereux) Chaloner. 663.17.

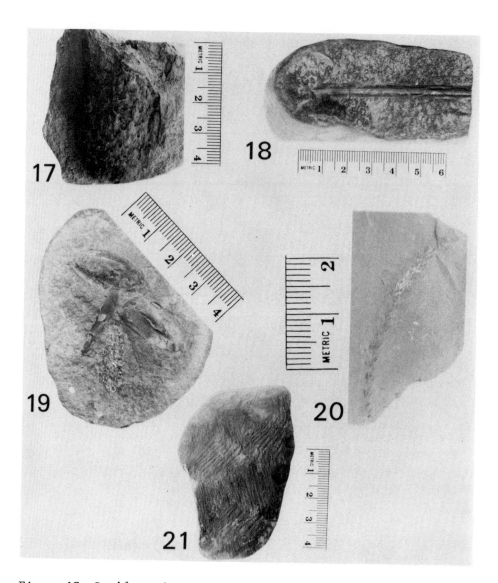

Figure 17. *Lepidostrobus ornatus* Brongniart. 663.112.
Figure 18. *Lepidophylloides* sp. Snigirevakaya. JJ 7.40.
Figure 19. *Lepidocystis* sp. Lesquereux. JJ 1.28.
Figure 20. *Lycopodites* sp. Lindley and Hutton. 663.108.
Figure 21. *Asolanus campotaenia* Wood. 538.

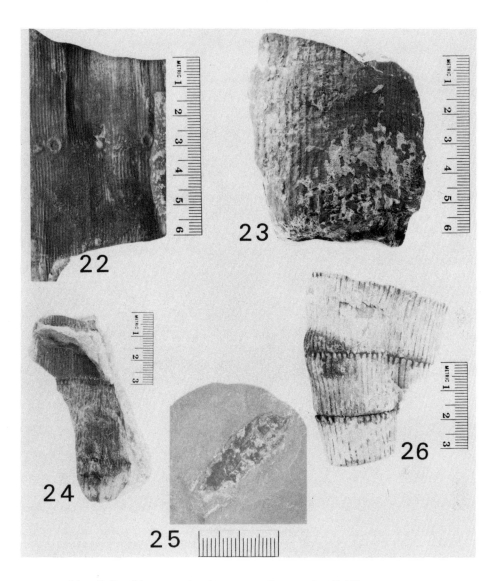

Figure 22. *Calamites carinatus* Sternberg. JJ 5.77.
Figure 23. *Calamites undulatus* Sternberg. 439.
Figure 24. *Calamites cistii* Brongniart. 315.
Figure 25. *Paracalamostachys cartervillii* Hibbert and Eggert. 663.184.
Figure 26. *Calamites suckowi* Brongniart. JJ 2.312.

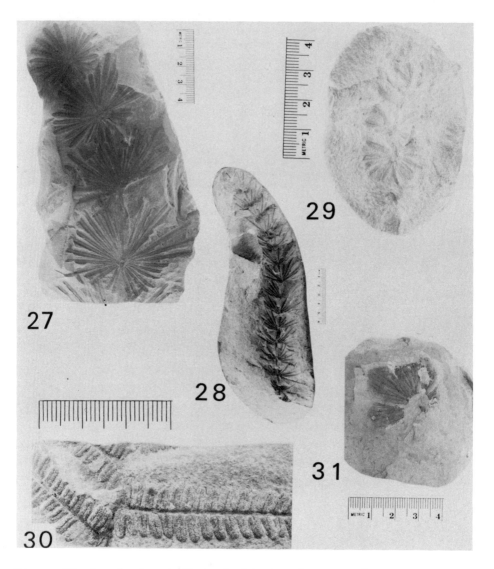

Figure 27. *Annularia stellata* (Schlotheim) Wood. 545.
Figure 28. *Asterophyllites equisetiformis* Schlotheim. 546.
Figure 29. *Annularia radiata* (**Brongniart**) Sternberg. JJ 2.163.
Figure 30. *Pecopteris plumosa* Artis (*Pecopteris dentata* Brongniart)
    JJ 14.2.
Figure 31. *Sphenophyllum* cf. *emarginatum* Brongniart. JJ 0.31.

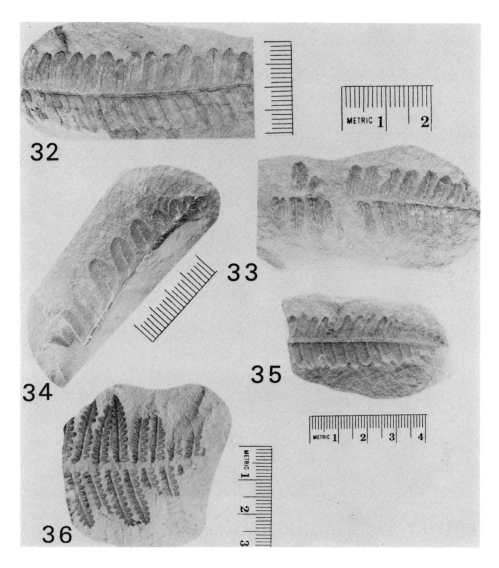

Figure 32. *Pecopteris bucklandi* Brongniart. 663.152.
Figure 33. *Pecopteris daubreii* Zeiller. 663.9.
Figure 34. *Pecopteris cisti* Brongniart. 441.
Figure 35. *Pecopteris* cf. *cyathea* Schlotheim. 539.
Figure 36. *Pecopteris arborescens* (Schlotheim) Brongniart. 485.

Figure 37. *Pecopteris* cf. *miltoni* Artis. JJ 1.24.
Figure 38. *Pecopteris lamuriana* Heer. JJ 4.2.
Figure 39. *Pecopteris oreopteridia* Schlotheim. JJ 6.1.
Figure 40. *Pecopteris psuedovestita* D. White. JJ 2.303.
Figure 41. *Pecopteris unita* Brongniart. 496.

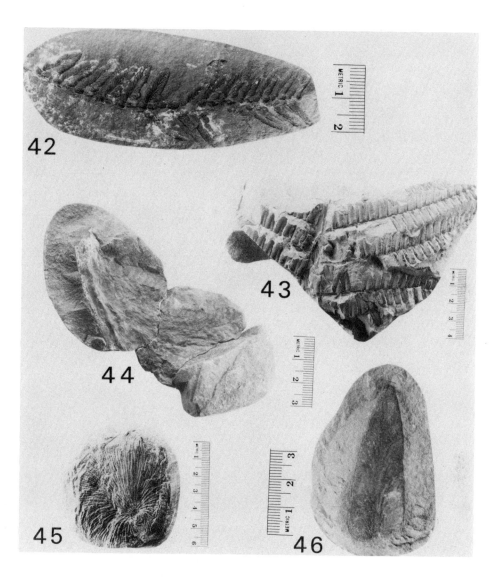

Figure 42. *Neuropteris heterophylla* Brongniart. JJ 2.61.
Figure 43. *Neuropteris rarinervis* Bunbury. 663.134.
Figure 44. *Neuropteris* cf. *missouriensis* Lesquereux. JJ 0.14.
Figure 45. *Cyclopteris trichomanoides* Brongniart. 2-1.2.
Figure 46. *Neuropteris anomala* Lesquereux. 663.121.

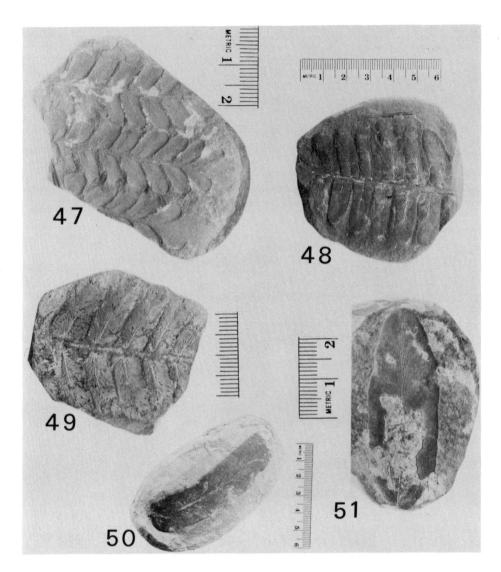

Figure 47. *Neuropteris ovata* Hoffman forma *typica* Crookall. 490.
Figure 48. *Neuropteris ovata* Hoffman forma *flexuosa* (Sternberg)
    Crookall. 474.
Figure 49. *Neuropteris ovata* Hoffman forma *vermicularis* (Lesquereux)
    Darrah. JJ 3.62.
Figure 50. *Neuropteris scheuchzeri* Hoffman forma *scheuchzeri* comb.
    nov. 663.32.
Figure 51. *Neuropteris scheuchzeri* Hoffman forma *decipiens* comb.
    nov. JJ 14.24.

Figure 52. *Renaultia chaerophylloides* (Brongniart) Zeiller. 442.
Figure 53. *Callipteridium rugosum* Lesquereux. JJ 12.60.
Figure 54. *Alethopteris serli* (Brongniart) Goeppert. 663.160.
Figure 55. *Alethopteris sullivanti* (Lesquereux) Schimper. 6-2.17.

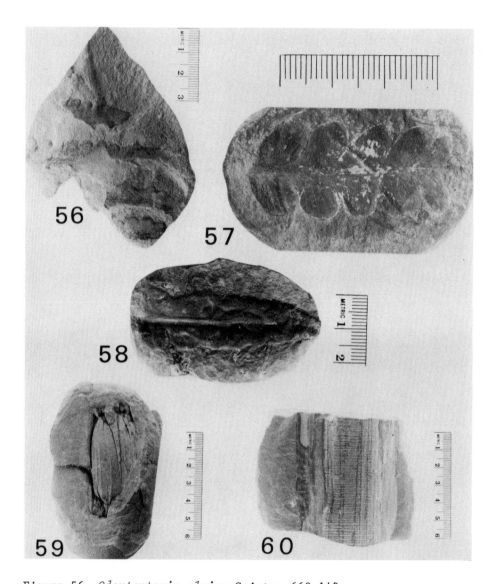

Figure 56. *Odontopteris alpina* Geintz. 663.140.
Figure 57. *Odontopteris* cf. *aequalis* Lesquereux. JJ 5.51.
Figure 58. *Odontopteris* cf. *abbreviata* Lesquereux. JJ 12.62.
Figure 59. *Trigonocarpus schutzianus* Goepp. and Berg. 663.191.
Figure 60. *Artisia* Sternberg. 534.

placed in synonymy with *Neuropteris scheuchzeri* due to the differences in the number of veins per centimeter of margin length, and
its highly localized occurrence. In order to indicate an affinity
of the two taxa, which may or may not be genetical, and still
separate them due to their consistent differences, *Neuropteris
decipiens* is placed as a separate form of *Neuropteris scheuchzeri*.

   Cyclopteris trichomanoides Brongniart has been identified, as
well as other Medullosan elements including *Odontopteris alpina*
Geintz, *Odontopteris* cf. *aequalis* Lesquereux, and *Odontopteris*
cf. *abbreviata* Lesquereux. The genus *Alethopteris* Sternberg is
represented by *Alethopteris serlii* (Brongniart) Goeppert and
Alethopteris sullivanti (Lesquereux) Schimper. *Callipteridium
rugosum* Lesquereux, a most confusing form as to taxonomic placement,
is also encountered. Reproductive organs of the Medullosaceae are
often noted but usually unidentifiable. One specimen assignable
to *Trigonocarpus schutzianus* Goepp. and Berg. is within the
collection.

   Lyginopterid elements are very infrequently found and when
encountered are usually poorly preserved. One species has been
identified, *Renaultia* (*Sphenopteris*) *chaerophylloides* (Brongniart)
Zeiller, and it is believed that these elements were probably
transported to the basin of deposition.

                           Coniferopsida

   There is an apparent absence of *Cordaites* at the Carterville
locality. Although *Artisia* Sternberg (Figure 60) is present,
Seward (1917) remarks that it is not reliable to assume that
specimens possessing a discoid type of pith are referable to the
Cordaitales because of the presence of transversal ridges in pith
casts of other genera.

                            DISCUSSION

   Twenty four genera and fifty two species compose the floral
assemblage from the strata above the Herrin (No. 6) Coal Member
at the Carterville locality. The primary constituents (Figure 2)
of the collection are the pteridophyll and pteridosperm foliage,
notably those assignable to *Pecopteris* Brongniart and *Neuropteris*
(Brongniart) Sternberg. Secondary elements of the assemblage
include foliage of the Sphenopsida, disarticulated sporophylls of
the Lepidodendrales, and other pteridosperm taxa. *Pecopteris
lamuriana* Heer, *Pecopteris unita* Brongniart, *Pecopteris cisti*
Brongniart, *Pecopteris plumosa* Artis (*Pecopteris dentata* Brongniart),
Neuropteris scheuchzeri Hoffman forma *scheuchzeri* comb. nov.,

Neuropteris scheuchzeri Hoffman forma *decipiens* (Lesquereux) comb.
nov., the *Neuropteris ovata* complex (*Neuropteris ovata* Hoffman
forma *typica* Crookall, *Neuropteris ovata* Hoffman forma *flexuosa*
(Sternberg) Crookall, and *Neuropteris ovata* Hoffman forma *vermicularis* (Lesquereux) Darrah), and *Neuropteris rarinervis* Bunbury
represent 48% of all specimens assignable to a specific rank.
The assemblage may be characterized as a *Pecopteris lamuriana -
Neuropteris scheuchzeri* forma *scheuchzeri - Neuropteris scheuchzeri*
forma *decipiens - Neuropteris ovata* complex association.

   Annularia stellata (Schlotheim) Wood, *Annularia radiata*
(Brongniart) Sternberg, and *Asterophyllites equisetiformis* Schlotheim
are quite abundant and may be considered as secondary components.
These forms are representative of 8% of the specific assemblage.
Sphenophyllum Brongniart is very rare at the site and was probably
transported into the basin. Disarticulated lycopsid sporophylls
are quite abundant although specimens representing vegetative organs
are infrequently encountered. This abundance of sporophylls may
be due to a number of factors including the transportability of
sporophylls as opposed to vegetative organs and selective preservation in a nodular state. Other secondary components include
Renaultia (*Sphenopteris*) *chaerophylloides* (Brongniart) Zeiller,
Alethopteris serli (Brongniart) Goeppert, *Alethopteris sullivanti*
(Lesquereux) Schimper, *Pecopteris bucklandi* Brongniart, *Pecopteris
daubreii* Zeiller, *Pecopteris arborescens* (Schlotheim) Brongniart,
and *Pecopteris oreopteridia* Schlotheim.

   Comparison of the Carterville assemblage with Mid-Continent
floras reveals that there is a great similarity to other Middle
Pennsylvanian localities. The most famous Middle Pennsylvanian
impression-compression flora in North America is that of the
Francis Creek Shale (Mazon Creek Flora equivalent to the Colchester
No. 2 Coal Member). Darrah (1969) states that there are about 200
nominal species encountered in the various collections, and that
75% of these are fern and pteridosperm foliage. The dominant
association from the Francis Creek Shale, composing 60% of the
flora, is represented by four taxa: *Neuropteris decipiens* (=
Neuropteris scheuchzeri forma *decipiens*), *Neuropteris ovata*,
Neuropteris rarinervis, and *Pecopteris lamuriana*. Five additional
species compose 20% of the Francis Creek Shale flora which are also
found as secondary components of the Carterville site. This
suggests that a majority of the forms of the Carterville assemblage
are a continuation of floristic elements within the Kewanee Group.
Forty-four taxa of the Carterville assemblage are also found within
the 149 taxa of the Carr and Daniels collection (as reported by
Stewart, 1950).

   The Middle Pennsylvanian of Missouri has been studied in quite
some detail (White, 1899; Basson, 1968). Of the 120 species
listed from the Des Moines Series Henry County Flora (White, 1899),

17 are found in the Carterville collection. There is a noticeable
absence of *Neuropteris ovata* in the Henry County flora, as well
as *Pecopteris lamuriana* and *Neuropteris decipiens* (= *Neuropteris
scheuchzeri* forma *decipiens*). Basson (1968) published a floristic
study of the Drywood Formation (Middle Des Moines Series) which
includes 21 taxa that are collected from the Carterville site.
There is again an absence of *Pecopteris lamuriana* and *Neuropteris
decipiens* in the flora. The absence of *Neuropteris decipiens* may
be due to an inclusion of this taxon with *Neuropteris scheuchzeri*
in the study of Basson, but White (1899) recognized a difference
between the two taxa and retained each as a separate species.

The contemporaneous systems in the Eastern United States with
the Kewanee Group are the Pottsville and Allegheny Series. Darrah
(1969) states that there is no floral break between these series
and the division between the two is arbitrary. Rather, there is
a transition of floristic elements within the Allegheny. The
Allegheny Series in the Appalachian Region is divided into the
Kittanning Coals (Lower, Middle, and Upper) and the Freeport
Coal. Darrah (1969) believes that the Middle-Upper Kittanning
Coal is an association of *Neuropteris ovata* - *Neuropteris
scheuchzeri* - *Pecopteris lamuriana* with satellite species of
Alethopteris serli and *Sphenophyllum emarginatum*. The association
of plants at the Carterville locality is comparatively similar,
though there is a greater diversification of pteridophyll and
pteridosperm foliage. The Freeport Coal is relatively analagous
in composition to the Kittanning Coals except for the introduction
of a greater diversity of *Pecopteris*. The Carterville assemblage
is being equated to the Upper Kittanning Coal rather than any
other stratigraphic unit based upon dominant plant association
comparisons and floral diversification.

ACKNOWLEDGEMENTS

Sincere thanks is expressed to Lawrence C. Matten for his
critical review of this manuscript and his advisement. I would
also like to thank William C. Darrah for his guidance, advice,
and moral support. James Jennings needs to be acknowledged for
his many years of collecting the Carterville material and depositing
the specimens in the SIU paleobotanical collections.

LITERATURE CITED

Abbott, M. L. 1958. The American Species of *Asterophyllites,
Annularia,* and *Sphenophyllum*. Bull. of Am. Paleont. 38(174):
289-390.
Basson, P. W. 1968. The fossil flora of the Drywood Formation of

southwestern Missouri. Univ. Mo. Stud. vol. 44. Univ. Mo.
    Press. Columbia, Mo.
Bertrand, P. 1930. Bassin houiller de la Sarre et de la Lorraine.
    I. Flore fossile le fasc. Neuropteridees. Etudes Gites Min.
    France. pp. 1-58.
Bertrand, P. 1932. Bassin houiller de la Sarre et de la Lorraine.
    I. Flore fossile 2me fasc. Alethopteridees. Etudes Gites Min.
    France. pp. 61-107.
Boersma, M. 1972. The Heterogeneity of the Form Genus *Mariopteris*
    Zeiller. Drukkerij. Elinkwijk. Utrecht.
Cady, R. 1947. Shipping Coal Mines of Illinois. Map. Illinois
    State Geol. Surv.
Chaloner, W. G. 1967. in E. Boureau, ed., Traite de Paleobotanique.
    vol. 2. Paris.
Corsin, P. 1932. Bassin houiller de la Sarre et de la Lorraine.
    I. Flore fossile 3e fasc. Mariopteridees. Etudes Gites Min.
    France. pp. 111-173.
Corsin, P. 1951. Bassin houiller de la Sarre et de la Lorraine.
    I. Flore fossile 4e fasc. Pecopteridees. Etude Gites Min.
    France. pp. 174-286.
Cridland, A., J. E. Morris, and R. W. Baxter. 1963. The
    Pennsylvanian plants of Kansas and their stratigraphic
    significance. Paleont. B 112:58-92.
Crookall, R. 1955. Fossil Plants of the Carboniferous Rocks of
    Great Britian. Great Britian Geol. Surv. Paleontol., Mem. 4,
    pt. 1. pp. 1-84.
Crookall, R. 1959. Fossil Plants of the Carboniferous Rocks of
    Great Britian. Great Britian Geol. Surv. Paleontol., Mem. 4,
    pt. 2. pp. 85-216.
Crookall, R. 1964. Fossil Plants of the Carboniferous Rocks of
    Great Britian. Great Britian Geol. Surv. Paleontol., Mem. 4,
    pt. 3. pp. 217-354.
Crookall, R. 1966. Fossil Plants of the Carboniferous Rocks of
    Great Britian. Great Britian Geol. Surv. Paleontol., Mem. 4,
    pt. 4. pp. 355-572.
Danze, J. 1956. Contribution a L'etude des *Sphenopteris* les
    Fougeres Sphenopteridiennes. Theses. Text and Atlas.
    Imprimeriedouriez-Battaille 5. Rue Jacquemans. Lille.
Darrah, W. C. 1969. Upper Pennsylvanian Floras of North America.
    Privately published.
Fontaine, W. F., and I. C. White. 1880. The Permian or Upper
    Carboniferous flora of West Virginia. Second Geol. Surv. Pa.
    vol. PP. pp. 143.
Gillespie, H. W., L. S. Latimer Jr., and J. A. Clendenning. 1966.
    Plant Fossils from West Virginia. West Va. Geol. Surv.
Hibbert, F. A., and D. A. Eggert. 1965. A new calamitalean cone
    from the Middle Pennsylvanian of southern Illinois. Paleon-
    tology. 8(4):681-684.
Janssen, R. 1939. Leaves and Stems from Fossil Forests. Ill.
    State Mus. Pop. Sci. Ser. vol. 1.

Janssen, R. 1940. Some Fossil Plant Types of Illinois. Ill.
    State Mus. Sci. Papers. I.
Jennings, J. R., and D. A. Eggert. 1972. *Senftenbergia* is not a
    Schizaceous fern. Amer. Jour. Bot. 59:676 (Abstract).
Kosanke, R. M., J. A. Simon, H. R. Wanless, and H. B. Willman.
    1960. Classification of the Pennsylvanian strata of Illinois.
    Ill. State Geol Surv. Rept. of Investigations. 214.
Langford, G. 1958. The Wilmington Coal Flora from a Pennsylvanian
    Deposit in Will County, Illinois. Esconi Associates.
Langford, G. 1963. The Wilmington Coal Fauna and Additions to
    the Wilmington Coal Flora from a Pennsylvanian Deposit in Will
    County, Illinois. Esconi Associates.
Lesquereux, L. 1879-1884. Description of the Coal Flora of the
    Carboniferous Formation in Pennsylvania and throughout the
    United States. Second Pennsylvania Geol. Surv. P. 3 vols.
    (Atlas, 1879; Text vols. 1-2, 1880; vol. 3, 1884).
Read, C. B., and S. H. Mamay. 1960. Upper Paleozoic floral zones
    and floral provinces of the United States. U. S. G. S. Prof.
    Paper 454-K.
Seward, A. C. 1917. Fossil Plants. vol. 3. 3rd edition. Hafner
    Publishing, New York.
Schopf, J. M. 1975. Modes of fossil preservation. Rev. Paleobot.
    Palynol. 20:27-53.
Stewart, W. N. 1950. Report on the Carr and Daniels Collection
    of Mazon Creek. Ill. Acad. Sci. Trans. 43:41-45.
Stidd, B. M. 1974. Evolutionary trends in the Marattiales.
    Ann. Mo. Bot. Gard. 61(2):388-407.
Wagner, G. 1962. A brief review of the stratigraphic and floral
    succession of the Carboniferous of northwestern Spain.
    4e Congr. Strat. Carb. C.R. 3:753-762.
White, D. 1899. Fossil Flora of the Lower Coal Measures of
    Missouri. U. S. G. S. Mon. 37.

PALEOBOTANICAL AND GEOLOGICAL INTERPRETATIONS OF PALEOENVIRONMENTS

OF THE EASTERN INTERIOR BASIN

Richard L. Leary

Illinois State Museum

Springfield, Illinois   62706

ABSTRACT

Several paleoenvironments are recognized on the basis of both
floral differences and lithologic variations. It is important that
paleoenvironments determined on paleobotanical grounds be correlated with those based upon geological considerations. A variety
of paleoenvironments are preserved in and around the Eastern Interior Basin, providing opportunities for gathering both geologic
and paleobotanical data.

INTRODUCTION

Several paleoenvironments are recognized on the basis of both
floral differences and lithologic variations. These include a
variety of lowland and "swamp" environments as well as moist and
dry upland situations. It is important that the correlations between paleoenvironments determined on paleobotanical grounds and
those based upon geological considerations be understood. Thus,
there is increasing need for additional data on the geological
disposition of plant fossils. This includes data on the sediments
enclosing the fossils, the associated sedimentary structures, and
paleotopography.

A variety of environments existed in and around the Eastern
Interior Basin during the Pennsylvanian Period, including extensive
"swamps," major valleys, tributary valleys, channels, sinkholes,
and caves. Extensive lowlands occupied large areas of the Basin
during much of the Pennsylvanian. Major valley systems developed

157

on Mississippian carbonates prior to Pennsylvanian sedimentation. During the Pennsylvanian, deltas, coastal marshes, channels, and other environments existed within the lowlands. On the margins of the Basin were both moist and dry upland areas. Valleys, channels, sinkholes, and other depressions developed on pre-Pennsylvanian strata exposed on the Basin margin.

Plant fossils, both macrofossils and microfossils, are known from examples of each of the topographic settings. However, additional collecting sites are required in each and more data needed on the paleofloras from each in order to separate differences resulting from time (evolutionary) and from location (environmental).

PALEOENVIRONMENTAL CONSIDERATIONS

In Europe, Havlena (1971) has delimited several Carboniferous environments on the basis of floral composition. The two major divisions are the hygrophilous and mesophilous floras. The former occurs in the humid environment of the coal basin and consists of three fundamental associations: Lepidophyta, Articulata, and Pteridophylla. The mesophilous floras existed either in places within the coal basin unfavorable for the hygrophilous flora or on the hills surrounding the basin. A third floral association recognized by Havlena (1971, p. 252) is the xerophilous flora. This plant assemblage existed on the drier uplands. Similar associations have been reported in the Eastern Interior Basin of the United States (Peppers and Pfefferkorn, 1970).

It is not enough to merely separate the floras on the basis of composition. Our understanding of these environment-influenced variations can only be expanded and refined by careful collecting and thorough recording of stratigraphic and lithologic data at each site.

In this regard, studies in the United States have detailed the existence of several environments during the Pennsylvanian, based upon lithologic data from well records. Wanless and his students (Wanless et al., 1963, 1969, and Wanless, 1969) produced a series of maps of the North American midcontinent showing eight environments which controlled coal accumulation: deltas, channels, estuaries, coastal marshes, lagoons behind offshore bars, meander cutoffs, plains exposed by sea regression and depositional plains.

There are still many unanswered questions; for example, Cordaites are found in swamp, offshore and upland associations. It is not known if they lived in all of these environments or if the tough, strap-like leaves floated from one environment to another.

During the early Pennsylvanian Period, upland floras existed on the rim of the Eastern Interior Basin, and probably elsewhere. The environment was undoubtedly different from that of the lowland within the Basin. The flora was definitely of a different composition.

Megalopteris, and some other genera, are believed to be upland
plants. However, little data has been recorded on the geological
characteristics of many collecting localities. Many publications
are restricted to a discussion of a single plant genus or species
from a locality with no complete list of associated species; others
consist of a description of the flora with no reference to the nature of the deposits.

Environmental or facies control of floral composition has significance beyond the availability of fossils of a particular genus.
Many plant species, and even genera, may occur only in limited environments. The presence of their fossils is facies controlled, and
hence they are found only rarely. This may give the impression of
a short stratigraphic range whereas the species or genus may have
endured over a long time in a limited environment.

Needless to say, this can play havoc with attempts to determine
the age of strata on the basis of a few plant species. It is, therefore, imperative that we determine which forms are restricted to narrow facies.

Another reason for expanding our knowledge of paleoenvironment
is that environment may affect the pace of evolution. In particular,
it has been suggested that the increased diversity within the upland
floras may have accelerated the pace of evolution (Axelrod, 1961).
It is also possible that the "more challenging environment," one with
greater temperature and moisture variation, may have brought about
more variability within the floras.

For all of the above reasons, and perhaps others, we need to
gather more data on ancient environments and to seek plant fossils
from more diverse environments.

REQUIRED DATA

The set of data which should be the easiest to provide is a
complete list of species present in a flora or at a locality. In
spite of this, many reports contain a description of a single form
or a small select group of fossils from a locality with no indication of the other forms present. Even a list of genera would help
to determine the possible flora at the site.

Even better than a list of plants present at a locality is a
determination of abundance of each species in terms of percentages
of the total flora. However, such numbers can be misleading and
their derivation needs to be explained. In some cases the number
of specimens are estimated in the field, carefully counted in the
field, or counted in the laboratory; each has its own built-in bias.

The lithologic and stratigraphic details of the deposit should
be recorded. Information on grain size, texture, composition and
the presence of sedimentary structures such as cross-bedding, graded-
bedding, ripple marks and mud cracks should be recorded. From these
data, the sedimentological environment in which the plants were deposited can be determined.

The age of the strata needs to be determined as accurately as
possible from stratigraphic evidence and field correlation. This
involves careful observation and comparison with known deposits
elsewhere.

PALEOTOPOGRAPHY

The following discussion is intended to give an indication of
some of the major types of deposits in the Midwest where one may en-
counter plant fossils. Each of these represents a slightly different environment and hence would be important as a plant-fossil collecting site.

Coal Swamp

One of the best-known environments, from the paleobotanical
point of view, is the coal swamp. As mentioned earlier, several
microenvironments are recognized within the lowland which extended
over much of the central United States. The stratigraphy and lithology of many of these horizons are described and mapped.
In many instances, the deposits containing the "swamp floras"
are too extensive to be delimited within a single exposure. However,
sufficient mapping of large-scale paleoenvironment has been done
(Wanless et al., 1963, 1969), such that exact identification of the
stratum may place the deposit in context. Observations of lithologic
characteristics of the fossil-bearing strata should be recorded in
order to aid in identification of microenvironments. Composition,
grain size, and the presence of sedimentary structures may indicate
former oxbows, floodplains, deltas, etc.

Major Valleys

Major pre-Pennsylvanian valleys have been reported from several
areas in the deeper part of the Illinois Basin (Smith, 1941; Potter
and Desborough, 1965; Bristol and Howard, 1971). The larger valleys
are incised as much as 450 feet below the peneplain surface and are
up to 20 miles wide (Bristol and Howard, 1971). Many of these valleys are steep sided and contain slump blocks of Mississippian limestone (Bristol and Howard, 1971). The valleys developed prior to
the beginning of Pennsylvanian sedimentation in the Basin and contain
some of the earliest Pennsylvanian-age sediments.
Many of the larger valleys are in the deeper, central part of
the Basin and are known primarily from drilling record data. They
are occasionally exposed at the margin of the Basin and should be examined for plant fossils.

Tributary Valleys

Smaller, tributary valleys are to be expected on the margins of the Illinois Basin, near the headwaters of major valleys. Several have been reported (e.g., Shaw and Gildersleeve, 1969; Leary, 1974a). They are more likely to contain fossils of non-lowland plants than are the sediments deposited in the deeper basin as it was filled and became an extensive lowland.

A small tributary valley or ravine is exposed in northeastern Brown County, Illinois (Leary, 1974b; Leary and Pfefferkorn, in press). The valley, 12 meters wide and 12 meters deep, was eroded in Mississippian limestone and dolomite. It was filled with silt and sand during early Pennsylvanian time and contains impressions of the plants which grew on the uplands adjacent to the valley. The sediments are irregularly bedded and were washed from the land areas on either side of the depression. The plant fossils occur almost exclusively in the coarser sediments, indicating that perhaps they were transported primarily during times of storms.

The flora contains numerous upland plants and few lowland genera (Leary, 1974b). This is consistent with the environment as evidenced by the topography of the unconformity surface. The upland plants existed around the ravine while the few lowland species probably lived in the moist area in the bottom of the ravine.

Channels

A nearly flat erosional surface developed on Devonian and Silurian carbonates in the northwest corner of the Illinois Basin. Sinkholes, channels and valleys occur on this surface and some depressions contain Pennsylvanian sediments (McGinnis and Heigold, 1974). Some of the early Pennsylvanian floras sampled in the late 1800's and early 1900's probably came from such isolated deposits (White, 1908).

Steep-sided channels were eroded in Devonian limestone in the Rock Island County area and filled with sediments during the early Pennsylvanian (Leary, 1974b, in press a). These channels are four meters deep and ten meters wide. The mudstone/shale filling the channels contains a diverse flora. The lithologic and paleotopographic evidence indicated that water-filled depressions occurred on the level upland surface. The paleobotanical evidence indicates a mixed dry upland (xerophilous) and moist lowland (hygrophilous) flora (Leary, 1974b, in press a). The water-filled channels apparently permitted the growth of hygrophilous elements on the otherwise dry upland.

Sinkholes

A variety of smaller solution and collapse features within the pre-Pennsylvanian erosion surface are known (e.g., Bretz, 1950). In

Illinois, sinkholes are exposed in quarries in Kankakee County (Ek-blaw, 1925) and Will County (Peppers, person communication). Spores and fragmentary plant fossils have been reported from some of these depressions (Peppers, personal communication). Isolated pockets of Pennsylvanian sediments have also been reported from subsurface data from Whiteside County, Illinois. These may represent deposition in sinkholes in a pre-Pennsylvanian erosional unconformity on the Silurian dolomite surface (McGinnis and Heigold, 1974).

Because karst development occurs in regions where valleys are entrenched well below the surrounding land surface underlain by jointed, soluble rocks, (Thornbury, 1969, p. 306), it indicates upland areas. Plant fossils from such areas might provide information on the nearby upland floras. However, caution must be used in interpreting the environment as the sediments and fossils are contemporaneous with a period of deposition, not formation of the depression. Even if filled while still in an upland situation, ponding may create, in the immediate vicinity of the depression, a moist microenvironment in an otherwise dry region.

Caves

Caves containing Pennsylvanian sediments have been observed in several parts of the Illinois Basin (e.g., northwest corner, Savage and Udden, 1921). A large cavern filled with Pennsylvanian-age silt and sand is exposed in a quarry south of St. Louis, Missouri (Leary, in press b). The cavern was 15 meters wide, 20 meters high, and at least 300 meters in length (Brill, 1973).

A few plant fossils occur within the silt and sand which fill the cave but preservation is poor. Cordaites is the only genus identified from the deposit although other forms have been observed. The large size of the Cordaites and the possible attachment of several leaves to a stem would indicate they had not been transported far and perhaps had entered the cave through a nearby sinkhole.

CONCLUSIONS

In addition to several climatic environments, a variety of depositional environments existed during the Pennsylvanian. Each provides us with its own particular set of possibilities and problems. In an effort to understand the fossil plant assemblages, determine accurately the stratigraphic ranges of species and follow the evolution of genera, we need to record geological data as well as systematically collecting plant fossils.

LITERATURE CITED

Axelrod, D.I., 1961. How old are the angiosperms? American Journal
    of Science, vol. 259, p. 447-459.
Bretz, G.H., 1950. Origin of the filled sink-structures and circle
    deposits of Missouri. Geological Society of America Bulletin
    61, p. 789-834.
Brill, K.G., 1973. Valley filled with clastic sedimentary rocks of
    Pennsylvanian age, St. Louis County, Missouri (Abs). Geologi-
    cal Society of America Abstracts with Programs 5(4), p. 302-303.
Bristol, H.M., and R.H. Howard, 1971. Paleogeologic map of the sub-
    Pennsylvanian Chesterian (Upper Mississippian) surface in the
    Illinois Basin. Illinois State Geological Survey Circular 458,
    16 p.
_____, 1974. Sub-Pennsylvanian valleys in the Chesterian surface
    of the Illinois Basin and related Chesterian slump blocks. In
    Carboniferous of the southeastern United States. Geological
    Society of America Special Paper 148, p. 315-336.
Ekblaw, G.E., 1925. Paleozoic karst topography. Illinois State
    Academy of Science Transactions, v. 17 (1924), p. 208-212.
Havlena, V., 1971. Die zeitgleichen Floren des Europaischen Ober-
    karbons und die Mesophile Flora des Ostrau-karwiner steinkohlen-
    reviers. Review of Palaeobotany and Palynology, 12(4), p. 245-
    270.
Leary, R.L., 1974a. Stratigraphy and floral characteristics of the
    basal Pennsylvanian strata in west-central Illinois. Compte
    Rendu, 7th International Congress on Stratigraphy and Geology
    of the Carboniferous, Krefeld, Germany, 1971, vol. 3, p. 341-
    350.
_____, 1974b. Two early Pennsylvanian floras of western Illinois.
    Illinois State Academy of Science Transactions, 67(4), p. 430-
    440.
_____, in press a. Early Pennsylvanian paleogeography of an upland
    area, western Illinois. Bulletin de la Society Belge de Geologie.
_____, in press b. Namurian paleogeography of the western margin
    of the Eastern Interior (Illinois) Basin. Compte Rendu, 8th
    International Congress on Stratigraphy and Geology of the Car-
    boniferous, Moscow, USSR, 1975.
Leary, R.L., and H.W. Pfefferkorn, in press. A lower Pennsylvanian
    flora with Megalopteris and Noeggerathiales from west-central
    Illinois. Illinois State Geological Survey Circular.
McGinnis, L.D., and P.C. Heigold, 1974. A seismic refraction survey
    of the Meredosia Channel Area of northwestern Illinois. Illinois
    State Geological Survey Circular 488, 19 p.
Peppers, R.A., and H.W. Pfefferkorn, 1970. A comparison of the floras
    of the Colchester (No. 2) Coal and Francis Creek Shale. In Dep-
    ositional environments in parts of the Carbondale Formation,
    western and northern Illinois. Illinois State Geological Survey
    Guidebook Series, No. 8, p. 61-74.

Potter, P.E., and G.A. Desborough, 1965. Pre-Pennsylvanian Evans-
    ville Valley and Caseyville (Pennsylvanian) sedimentation in the
    Illinois Basin. Illinois State Geological Survey Circular 384,
    16 p.
Savage, T.E., and J.A. Udden, 1921. The geology and mineral resources
    of the Edgington and Milan quadrangles. Illinois State Geologi-
    cal Survey Bulletin 38C, 96 p.
Shaw, F.R., and B. Gildersleeve, 1969. An anastomosing channel com-
    plex at the base of the Pennsylvanian system in western Kentucky.
    U.S. Geological Survey Professional Paper 650-D, p. D206-D209.
Smith, M.H., 1941. Structure contour map of the pre-Pennsylvanian
    surface of Illinois. Illinois State Academy of Science Trans-
    actions, 34(2), p. 160-163.
Wanless, H.R., 1969. Marine and non-marine facies of the Upper Car-
    boniferous of North America. Compte Rendu, 6th International
    Congress on Stratigraphy and Geology of the Carboniferous, Shef-
    field, England, 1967, vol. 1, p. 293-336.
Wanless, H.R., J.B. Tubb, Jr., D.E. Gednetz and J.L. Weiner, 1963.
    Mapping sedimentary environments of Pennsylvanian cycles. Geo-
    logical Society of America Bulletin 74, p. 437-486.
Wanless, H.R., J.R. Baroffio and P.C. Trescott, 1969. Conditions of
    deposition of Pennsylvanian coal beds. In Environments of coal
    deposition, E.C. Dapples and M.E. Hopkins, editors, Geological
    Society of America Special Paper 114, p. 105-142.
White, D., 1908. Report on field work done in 1907. In Illinois
    State Geological Survey Bulletin 8, p. 268-272.

PRELIMINARY INVESTIGATION OF TWO LATE TRIASSIC CONIFERS FROM YORK COUNTY, PENNSYLVANIA

Bruce Cornet

Department of Geosciences, The Pennsylvania State University, University Park, Pennsylvania 16802

ABSTRACT

Two conifers are briefly described and illustrated from a rich plant-bearing layer of late Carnian age from the middle New Oxford Fm., Gettysburg Basin, USA. Two types of leaf-bearing branch systems are assigned to *Pagiophyllum diffusum* (Emmons) nov. comb. and *P. simpsonii* Ash. Seed cones that I assign to *Glyptolepis* nov. sp.1 and cf. *G.* nov. sp. 2 were also found. Six cones of *G.* nov. sp. 1, bearing large ovoid winged seeds, were found attached to a branch bearing leaves of *P. diffusum*. Both types of leaves and cones possess the same hairy type of cuticle. A small, elongate pollen cone, also with a hairy cuticle, was recovered from the plant layer. This cone contains *Patinasporites densus* Leschik pollen. *P. densus* and a morphologically similar pollen type, *Vallasporites ignacii* Leschik, dominate the palynoflorule from the plant layer.

During the summer of 1975 a rich fossil plant layer, about 15 cm thick, was discovered in a tributary of Little Conewago Creek, south-west of York Haven, Pa., in the Gettysburg Basin. The plant layer belongs to the middle New Oxford Formation, and has been palynologically dated by me as late Carnian. Wanner and Fontaine (1900) described this flora from nearby excavation pits, although they misidentified many of the conifers, even describing as *Baiera muensteriana* an ovuliferous cone I recognize as cf. *Glyptolepis* nov. sp. 3. The new collection includes pinwheels of *Dinophyton* nov. sp., a genus previously reported only from the Late Triassic (mostly Carnian) Chinle Formation by Sidney R. Ash (1970a). My species of *Dinophyton* differs from *D. spinosus* Ash by lacking epidermal hairs.

165

Figure 1.   Ovuliferous cone scale of *Glyptolepis* nov. sp. 1, ca.
3 mm wide; redrawn from one of scales (arrows) in Figure 6.

Preliminary study of fragments of large branch systems (up to
three branch orders) of two species of *Pagiophyllum*, occasionally
with numerous attached ovuliferous cones (Fig. 8), reveals the presence of *Glyptolepis* Schimper 1872, a genus for seed cone common to
the Triassic of Western Europe. Although the cuticles of each species are superficially very similar, both being densely covered with
short and long epidermal hairs (generally 15-180 *mu* in length), the
ultimate leaves are distinct, having been referred to *Palissya* and
Cheirolepis by Wanner and Fontaine (1900).

Glyptolepis nov. sp. 1 (Figs. 6-8) appears to belong to the
"*Palissya*-like" vegetative branches (Fig. 3) of *Pagiophyllum diffusum* (Emmons) nov. comb. (basionym: *Walchia diffusus* Emmons 1856, p.
105, Pl. 3, fig. 2), based on one leafy shoot of *P. diffusum* attached to the cone-bearing branch in Figure 8 (arrow). The ovuliferous
scales of *Glyptolepis* nov. sp. 1 resemble those of *Glyptolepis keuperiana* Schimper (see Mägdefrau, 1963), but each scale has only 3-4,
narrow median lobes (Fig. 1) instead of 5-7 lobes, and bears two,
large, ovoid winged seeds (Figs. 5-7). The seeds of *G. keuperiana*
apparently lack wings.

The cones of cf. *Glyptolepis* nov. sp. 2 (Fig. 2) are less common, as are its presumed "*Cheirolepis*-like" vegetative branches (Fig.
4), which appear to be identical to *Pagiophyllum simpsonii* Ash from
the Chinle Fm. of the south-western United States (see Ash, 1970b).
Because no leafy shoots are borne on the only branch found that bears
cones of cf. *Glyptolepis* nov. sp. 2, my assignment of *P. simpsonii*
to the second species of *Glyptolepis* is based soley on the fact that
no other species of leafy shoot or vegetative branch with epidermal
hairs has been found in the flora: The flora is represented by a
collection of more than one hundred specimens, including one additional species of conifer with papillate cuticle (one, large, attached terminal cone bears 12 mm wide ovuliferous scales of cf. *Glyptolepis* nov. sp. 3 - not illustrated), leafy shoots of which bear a
strong resemblance to specimens of "*Palissya braunii*" in Fontaine
(1883) and to "*P. sphenolepis*" in Wanner and Fontaine (1900).

Figure 2. Ovuliferous cone scale of cf. *Glyptolepis* nov. sp. 2,
3.2 mm wide; a reconstruction based on transfer preparations of
several fragments of cone scales from three cones attached to the
same branch.

The ovuliferous scales of cf. *Glyptolepis* nov. sp. 2 (Fig. 2)
differ from *G.* nov. sp. 1 by having two large lateral lobes instead
of one, and about five or less? median lobes of irregular size, as
well as a bract which projects above the ovuliferous scale instead
of being basal as in *G.* nov. sp. 1. Because seeds have not been
found in cones of cf. *G.* nov. sp. 2, the number of seeds per scale

Figure 3. *Pagiophyllum diffusum* (Emmons) nov. comb., showing flat
leaves with a slightly constricted base (2.5-4.5 mm in overall length),
oriented in a single plane but spirally attached as in *Tsuga canadensis*; X 1.5.

Figure 4. *Pagiophyllum simpsonii* Ash, showing three-dimensional leaves with a slightly constricted base and a median keel (2-3 mm in overall length), spirally oriented and attached; X 1.4.

is not known, and a provisional assignment (cf.) to *Glyptolepis* is necessary.

Two overlapping fragments of one?, narrow, elongate pollen cone, the longest fragment of which is 1.8 cm in length, were found associated with the flora (Fig. 10). The microsporophylls possess the same type of hairy cuticle as *Pagiophyllum diffusum* and *P. simpsonii*. These fragments of cone possess numerous undehisced pollen sacs containing (poorly preserved) pollen grossly indistinguishable from *Patinasporites densus* Leschik (Fig. 9). The microsporophyll axes are relatively long and narrow (as is the cone axis), and appear to attach at the apical pointed end of the microsporophyll head. Two to four pollen sacs appear to fuse just before an expanded common base broadly attaches to the head at the point of attachment of the axis. The basal part of the head is enlarged to the point of shielding the pollen sacs.

Patinasporites densus and a morphologically similar pollen, *Vallasporites ignacii* Leschik, dominate the palynoflorule (60+%). Because two similar species of *Glyptolepis* and/or *Pagiophyllum* dominate the megafossil flora, it is tempting to consider *V. ignacii*

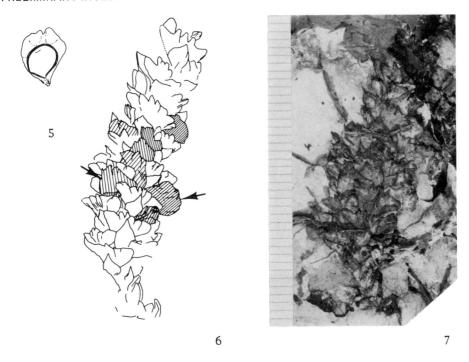

Figure 5.   Isolated seed identical to that found in cones of *Glyptolepis* nov. sp. 1; camera lucida drawing; 6 mm wide.

Figure 6.   Ovuliferous cone of *Glyptolepis* nov. sp. 1 containing at least nine seeds, borne in pairs (arrows) and laterally positioned relative to ovuliferous scales; camera lucida drawing of specimen in Fig. 7; seeds cross hatched.

Figure 7.   Transfer preparation of an isolated ovuliferous cone of *Glyptolepis* nov. sp. 1, showing a short leafy shoot at base; scale in mm.

as being produced by one of these two conifers, as is *P. densus*. However, these two species of pollen occur together in the Carnian of Europe where other species of *Glyptolepis* are present.   Therefore, it is probable that *P. densus* (and perhaps also *V. ignacii*) does not indicate the presence of any particular species of *Glyptolepis* at a regional level.   In other words, the *Patinasporites/Vallasporites* complex may have been produced by a number of species of conifer, including *Glyptolepis*, which would explain the prevalence of these two pollen types in the Late Triassic, particularly the Carnian.

The palynoflorule also contains one additional, but rare, species of *Patinasporites*, which is considerably larger than *P. densus*.

Figure 8. A branch bearing at least six ovuliferous cones of *Glyptolepis* nov. sp. 1; some cones still possess one or more seeds;
each cone is attached to the branch by means of a short leafy shoot;
cones are attached at various points on the branch, and because of
different orientation during burial, some cones became distorted
(shortened) due to compression; an ultimate leafy shoot identical
to *Pagiophyllum diffusum* arises near the base of one cone (arrow) -
counterpart demonstrates attachment of shoot, but contains less of
the cones; X 1.4.

Also present are *Alisporites parvus* de Jersey (common), *Triadispora*
spp., *Enzonalasporites vigens* Leschik, *Pseudoenzonalasporites sum-
mus* Scheuring, and possibly *Camerosporites pseudoverrucatus* Scheuring.

A late Carnian age is mostly based on the abundance of *P. densus* and *V. ignacii* (ca. 50% and 10% respectively), and on a well-
developed middle Carnian palynoflora in older strata of the Gettysburg, Taylorsville, Richmond, and Deep River Basins (cf. Dunay and
Fisher, 1974). Correlation of the Little Conewago Creek plant locality with the (lower) Lockatong Formation of the Newark Basin is
based on the presence of the fish, *Turseodus* sp., in a lacustrine
unit about 6 m above the plant layer (P. Olsen, pers. comm., 1975).
The stratigraphic range of *Turseodus* in the Newark Basin appears to
be limited to the Lockatong Fm.

ACKNOWLEDGEMENTS

The author acknowledges the support of National Science Foundation grant number GA-36870 to Professor Alfred Traverse, Department

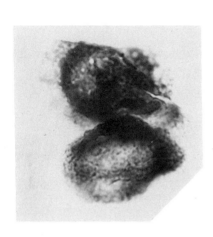

9

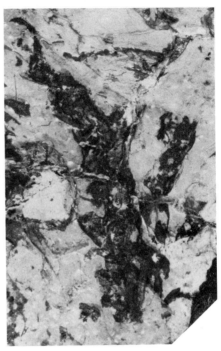

10

Figure 9. Three specimens of *Patinasporites densus* pollen isolated from the larger of the two pollen-cone fragments in Fig. 10; upper left specimen with expanded circumequatorial saccus is 53 *mu* in diameter, while the lower specimen showing a less expanded equatorial saccus is 52 *mu* in diameter.

Figure 10. Transfer preparation of two overlapping fragments of perhaps one pollen cone, which appears to belong to either *Glyptolepis* nov. sp. 1 or cf. *G.* nov. sp. 2, based on the presence of a similar hairy cuticle; a winged seed of the *G.* nov. sp. 1 type occurs just to the left of the cone fragments (white center); X 4.2.

of Geosciences, The Pennsylvania State University, for support of this study. I also acknowledge the helpful suggestions of and comparative material provided by Professor Sidney R. Ash, Department of Geology and Geography, Weber State College. Professor A. Traverse is further acknowledged for his assistance in preparing and criticizing this manuscript. Paul E. Olsen, Yale University, is acknowledged for information on the fishes and for numerous discussions about evidence for stratigraphic correlation. I also acknowledge my wife, Ginny, for her help in collecting the plant specimens used in this study, and Tate Ames, P.S.U., for suggested improvements in the manuscript.

REFERENCES

Ash, S. R.   1970a. *Dinophyton*, a problematical new plant genus from
     the Upper Triassic of the south-western United States. Palaeont.
     13: 646-663.

_____. 1970b. *Pagiophyllum simpsonii*, a new conifer from the Chin-
     le Formation (Upper Triassic) of Arizona. Journ. Paleont. 44:
     945-952.

Dunay, R. E. and M. J. Fisher. 1974. Late Triassic palynofloras of
     North America and their European correlatives. Rev. Palaeobot.
     and Palyn. 17: 179-186.

Emmons, E. 1856. Geological report of the midland counties of North
     Carolina. North Carolina Geol. Surv. 1852-1863: G. P. Putnam
     and Co., New York; Raleigh, H. D. Turner.

Fontaine, Wm. M. 1883. Contributions to the knowledge of the older
     Mesozoic flora of Virginia. U. S. Geol. Surv., Monogr. 6: 1-
     144.

Mägdefrau, K. 1963. Die Gattungen *Voltzia* und *Glyptolepis* im Mit-
     tleren Keuper von Hassfurt (Main). Geol. Bl. NO-Bayern 13:
     95-98.

Schimper, W. Ph. 1870-1872. Traité de paléontologie végétale 2
     and Atlas, Paris.

Wanner, H. E. and Wm. M. Fontaine. *in* Ward, L. F. 1900. Status of
     the Mesozoic floras of the United States. U. S. Geol. Surv.,
     20th Ann. Rept. 2: 233-315.

NORTH AMERICAN PRIMITIVE PALEOZOIC CHAROPHYTES AND DESCENDENTS

James E. Conkin[1] and Barbara M. Conkin[2]

University of Louisville[1] and Jefferson Community College[2]

Louisville, KY. 40208 and 40202

ABSTRACT

North American primitive Paleozoic charophytes consist of three orders: Chovanellales (proposed herein), Sycidiales, and Trochiliscales, disposed in three families (Chovanellaceae, Syciaceae, and Trochiliscaceae) and three genera, *Chovanella*, *Sycidium*, and *Moellerina* with seven species generically allocated as follows: *Chovanella* (1), *Moellerina* (4), and *Sycidium* (2). The fourth, and evolutionarily advanced order of charophytes, Charales, is represented in the Paleozoic by the sinistrally spiralled Middle Devonian *Eochara* (Family Eocharaceae) which gave rise to the Pennsylvanian *Palaeochara* (Palaeocharaceae) which subsequently gave rise to the Stellatocharaceae, new family, the ancestral stock of the clavatoraceids, and through the clavatoraceids to the lagynophoraceids. The porocharaceids, also derived from *Eochara*, are the stem from which the Mesozoic and Cenozoic charoideid and nitelloideid charophytes evolved.

A new family, Stellatocharaceae, is proposed to include those genera previously placed in the subfamily Stellatocharoideae (non-utricled, but with apical beak). The gyrogonite of *Catillochara* Peck and Eyer, 1963 has heretofore been inverted, and it is herein considered to be a junior synonym of *Stomochara* Grambast, 1961. *Stomochara* is placed in the family Porocharaceae, for not only is the apical pole of *Stomochara* different from that of *Stellatochara* Horn af Rantzien, 1954, but also the shapes of the gyrogonites and the geologic ranges of the two genera are different: *Stomochara* ranged from the Pennsylvanian to Triassic, but *Stellatochara* is a Mesozoic form known from the Triassic.

173

A fundamental dichotomy of the porocharaceid stock into nitel-loideids (with a double tier of crown cells) and charoideids (char-eids) (with a single tier of crown cells) occurred in the Late Jur-assic and both groups became co-dominant in the Quaternary. The raskyellaceids (Late Cretaceous-Late Oligocene) branched from the chareid stock in the Late Cretaceous; these forms close their large apical pore by means of an apical operculum (surrounded by a circu-lar dihescent suture). The gyrogonids (with peripheral modifica-tions and nodose ornamentation of the apical zones) branch from the chareid stock in the Early Paleocene and persisted until the Late Oligocene. The chareid characeans persist until the present day, and are probably represented in the Late Eocene by *Chara* (or *Char-ites*, the organ genus in which the gyrogonite is indistinguishable from *Chara*). In the chareids there is progressive diminution (and finally essential disappearance) of the apical opening.

The scheme of orientation of the gyrogonites presented by Conkin, *et al.* (1970, 1972, and 1974) is consistent with the orien-tation of oogonia of modern charophytes and has resulted in the suppression of *Karpinskya* (Croft, 1952) Grambast, 1962, the correct orientation of Devonian and/or Mississippian *Moellerina, Chovanella,* and *Sycidium*, as well as a number of Pennsylvanian, Permian, and Mesozoic genera, and clarification of evolution within the charo-phytes. An evolutionary lineage within the genus *Moellerina* Ulrich, 1886, emend. Conkin, *et al.*, 1974, based on progressive prominence of the cage superstructure, permits recognition of charophyte zones within the Middle and Upper Devonian and Lower Mississippian. Fur-ther, the usefulness of *Moellerina greenei* for precise age deter-mination and correlation within the Middle Devonian of eastern North America is demonstrated.

INTRODUCTION

Peck (1934, 1946, and 1953) and Peck and Morales (1966) dis-cussed the history of the study of charophytes and this need not be repeated here. Presently, we are concerned with the proper orienta-tion of the gyrogonites of fossil charophytes, for their correct orientation greatly assists in clarifying their taxonomy and their phylogenetic history. Emphasis is also placed on the stratigraphic value of charophytes.

ECOLOGY AND PRESERVATION

Charophytes, reproducing sexually, represent the highest class of algae, and while generally dioecious, are sometimes monoecious. Oogonia calcify (Text-fig. 1A) and thus frequently becomes fossil-ized gyrogonites (Text-fig. 1B); antheridia (Text-fig. 1C) are

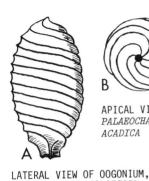

APICAL VIEW OF
*PALAEOCHARA
ACADICA*

ANTHERIDIUM OF
MODERN CHAROPHYTE

LATERAL VIEW OF OOGONIUM,
WITH SLIGHTLY CALCIFIED
ENVELOPING CELLS, SHOWING
CAGE SUPERSTRUCTURE AT BASE

VEGETATIVE CORTICATED
"STEM" OF *SYCIDIUM FOVEATUM*
(AFTER CONKIN, *ET AL.*, 1970)

CAGE SUPERSTRUCTURE
OF *SYCIDIUM FOVEATUM*
(AFTER CONKIN, *ET AL.*, 1970)

LATERAL VIEW OF
SPORANGIUM OF *CHARA*
SP., SHOWING
BASAL "CLAW"

LATERAL VIEW OF SPORANGIUM
OF <u>CHARA</u> SP., SHOWING BASAL CAGE

OBLIQUE LATERAL-BASAL
VIEW OF SPORANGIUM OF
CHARA SP. SHOWING TH
BASAL CAGE

OBLIQUE LATERAL-
BASAL VIEW OF
OOSPORANGIUM
SHOWING BASAL SCAR

LATERAL VIEW
OF CHAREID
OOGONIUM SHOW-
ING ONE-TIERED
CORONULA

LATERAL VIEW OF
NITELLEID OOGONIUM
SHOWING TWO-TIERED
CORONULA

LATERAL CROSS SECTION
OF OOGONIUM SHOWING
OOSPORE, OOSPORANGIAL
WALL, ENVELOPING CELL,
AND BASAL PLUG

TEXT-FIGURE 1. MORPHOLOGIC FEATURES OF MODERN AND FOSSIL
CHAROPHYTES.

rarely found as fossils and none is substantiated from the Paleozoic.
We have recovered a few antheridia from Pleistocene lake bed deposits, near Estelline, Hall County, Texas. Antheridial leaves of
Clavator reidi (Harris, 1939, pl. 1, figs. 2, 4, 6) and of *Perimneste
horrida* (Harris, 1939, pl. 14, figs. 4, 5) have been reported from
the Upper Jurassic Purbeck beds of England. Vegetative parts doubt-
fully referred to the charophytes are reported from the Devonian
Rhynie chert of Scotland (Kidston and Lang, 1921). Undoubted corticated "stems" (Text-fig. 1D) and cage superstructure (Text-fig. 1E)
of *Sycidium foveatum* Peck, 1934 are present in the Upper Devonian
Lime Creek Formation of Iowa (Conkin, *et al.*, 1972). Vegetative
parts of charophytes are found in the Mesozoic and are locally
abundant in Quaternary lake bed deposits.

Charophytes usually inhabit shallow, still to gently running
fresh water, but they can live in brackish water and in water depths
in some excess of 20 feet. Charophytes can withstand salinities up
to 26 parts/1000, can tolerate a wide range of temperatures, and
are essentially world wide in their geographic distribution. The
pH of water inhabited by charophytes may range from 5.0-9.6+, but
usually it is alkaline. Charophytes are in instances (notably in
the Middle Devonian, Lower Mississippian, Pennsylvanian, and Eocene)
associated with sediments and faunas of shallow marine environments
and, in some of these cases, occur above paracontinuities (Conkin
and Conkin, 1973a and 1975a).

ORIENTATION OF THE GYROGONITE

The gyrogonites of fossil charophytes are oriented in the same
manner as the oogonia of modern charophytes; criteria for determining the apical and basal poles of oogonia and gyrogonites are
detailed in Table 1.

In living charophytes, the cage (if present) is at the base of
the oosporangium (Text-figs. 1F-H) and consists of solid rods of
organic material which extend downward from the base of the oosporangium; a thin connecting veneer of organic material overlaps and
sags into the spaces between the rods. The cage is overlapped in
turn by the weakly calcified basal portion of the gelatinous spiral
enveloping cells. The thinly calcified basal portion of the spiral
cells over the cage is the cage superstructure (Text-fig. 1A) which
permits easy determination of the base of an oogonium (or a fossil
gyrogonite). Further, there is a rather large basal attachment
scar (Text-fig. 1I) at the base of the oogonium (or gyrogonite),
usually of a pentagonal form, and a basal plug which is surrounded
by the cage superstructure (Text-fig. 1L). Some of these structures
are absent in particular groups of charophytes; however, all charophyte oogonia (or gyrogonites) possess some combination of these

APICAL POLE	BASAL POLE
GENERAL OUTLINE OF THE APICAL END IS MORE REGULARLY ROUNDED AND OBTUSELY ANGLED[1]	GENERAL FORM OF THE BASAL END IS LESS ROUNDED AND ACUTELY ANGLED[1]
SMALL TO MINUTE OR ESSENTIALLY NO APICAL PORE IN OOGONIA; SMALL APICAL PORE IN SOME GYROGONITES	MEDIUM TO LARGE BASAL PORE IN BOTH OOGONIA AND GYRO-GONITES
ENVELOPING CELLS MEET AT NEARLY A POINT SURROUNDING A SMALL APICAL PORE (IN GYRO-GONITES) OR JOIN AT THE APEX TO FORM A THIN, SHORT, STRAIGHT OR ZIGZAG RIDGE WITH A MINUTE APICAL PORE (OR ESSENTIALLY NO PORE) AT THE CENTER (IN OOGONIA)	ENVELOPING CELLS ARE TRUNCA-TED BY THE BASAL PORE (POSI-TION OF THE BASAL ATTACHMENT SCAR) AND CAGE SUPERSTRUCTURE
POLYGONS (IN THE GYROGONITE OF *SYCIDIUM*) BECOME SMALLER AND MORE IRREGULAR AS THEY GRADUALLY LOSE SOME OF THEIR SIDES IN APPROACHING THE APICAL END	POLYGONS (IN THE GYROGONITE OF *SYCIDIUM*) ARE SOMEWHAT LARGER, RETAIN MORE SIDES AND MORE OF THEIR REGULARITY AS THE BASAL PORE IS APPROACHED
CORONULA IN A SMALL CIRCLET UNIT IN THE CENTER OF APICAL END; CORONULA NOT KNOWN TO BE CALCIFIED IN PALEOZOIC FORMS; THE CROWN CELLS ARE CUT OFF FROM THE ENVELOPING CELLS, BUT ARE IN THE SAME LINE; NO APICAL PLUG	CAGE IS AN EXTENSION OF THE BASAL PART OF THE OOSPORANIGAL WALL AND ENCLOSES A LARGE BASAL PLUG; THE CAGE IS OVERLAPPED BY THE CAGE SUPERSTRUCTURE WHICH IS AN EXTENSION OF THE ENVELOPING CELLS
NO INDENTATION OF THE NORMALLY CALCIFIED ENVELOPING CELLS AT THE APICAL END (THIS DOES NOT REFER TO THE CROWN CELLS OR TO OTHER CORONULAR STRUCTURES)	WALLS OF THE CAGE ARE SOMEWHAT CONCAVE WHICH CAUSE DEPRESSIONS OR PIT-LIKE SCARS ("SAGS") WHERE THE CAGE SUPERSTRUCTURE OVERLAPS THE CAGE
NO ATTACHMENT SCAR ON THE APICAL END	ATTACHMENT SCAR ON THE BASAL END

TABLE 1. APICAL AND BASAL FEATURES OF CHAROPHYTE OOGONIA AND FOSSIL GYROGONITES (AFTER CONKIN, *ET AL.*, 1974).

1. A POSSIBLE EXCEPTION TO THIS IS THE STELLATOCHARACEIDS, THE CLAVATORACEIDS, AND THE LAGYNOPHORACEIDS IN WHICH THE APICAL END HAS BEEN CONSIDERED TO BE LESS ROUNDED, AND ACUTELY ANGLED.

characteristic basal structures. Text-figure 2 illustrates the
orientation of selected genera within the four orders of charophytes.

   Past workers since Peck (1934), except Conkin, *et al.* (1970,
1972, and 1974) have misinterpreted the basal structure, that is,
cage superstructure as an apical structure, the coronula (crown
cells). As an example, Conkin, *et al.* (1974) already have shown
the cage superstructure in *Moellerina* has been misinterpreted as a
coronula; indeed, it was the supposed presence of the coronula that
was the basis for Croft's (1952) division of the genus *Trochiliscus*
Karpinsky, 1906 into two subgenera: *Karpinskya*, with supposed coronula and *Eotrochiliscus* (*Trochiliscus*) without coronula. Even
though the cage superstructure in these forms has been interpreted
in the past as a coronula, the cage superstructures resemble it
only superficially, for in living charophytes the coronula constitutes a discrete single (Text-fig. 1J) or double tier (Text-fig.
1K) of non-calcifying crown cells located at the center of the apical end of the oogonium, closely surrounding a very small to microscopic apical pore, and although separated from the enveloping cells,
they are lined up with them. Crown cells are not known to be preserved in any Paleozoic gyrogonites.

CLASSIFICATION OF AMERICAN PALEOZOIC CHAROPHYTES

Order Chovanellales Conkin and Conkin, new order
  Family Chovanellaceae Grambast, 1962 emend.
    Genus *Chovanella* Reitlinger and Yartseva, 1958 emend.
    Chovanella burgessi Peck and Eyer, 1963
Order Sycidiales Peck, 1934
  Family Sycidiaceae Peck, 1934
    Genus *Sycidium* Sandberger, 1849
    Sycidium calthratum Peck, 1934
    S. foveatum Peck, 1934
Order Trochiliscales Mädler, 1963
  Family Trochiliscaceae Peck, 1934
    Genus *Moellerina* Ulrich, 1886 emend. Conkin, *et al.*, 1974
    Moellerina greenei Ulrich, 1886 emend. Conkin, *et al.*, 1974
    M. bilineata (Peck), 1934 emend. Conkin, *et al.*, 1974
    M. laticostata (Peck), 1934 emend. Conkin, *et al.*, 1974
    M. convoluta (Peck), 1936, Peck and Morales, 1966
Order Charales
  Family Eocharaceae Grambast, 1959
    Genus *Eochara* Choquette, 1956
    Eochara wickendeni Choquette, 1956
  Family Palaeocharaceae Pia, 1927
    Genus *Palaeochara* Bell, 1922
    Palaeochara acadica Bell, 1922
  Family Porocharaceae Grambast, 1962

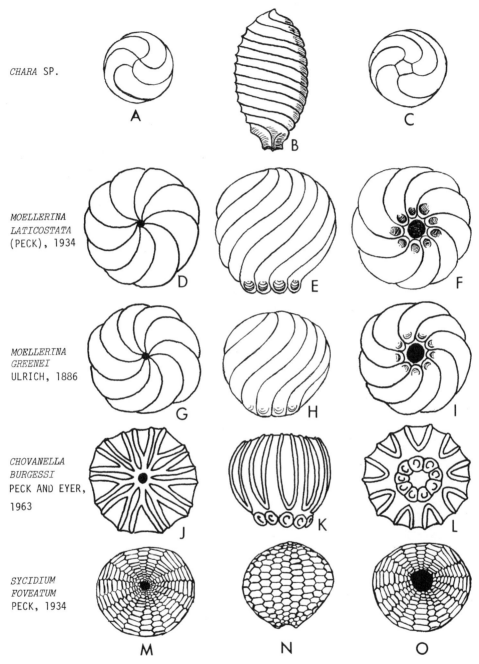

CHARA SP.

MOELLERINA
LATICOSTATA
(PECK), 1934

MOELLERINA
GREENEI
ULRICH, 1886

CHOVANELLA
BURGESSI
PECK AND EYER,
1963

SYCIDIUM
FOVEATUM
PECK, 1934

TEXT-FIGURE 2. ORIENTATION: CHARALES (A-C); TROCHILISCALES (D-I);
CHOVANELLALES (J-L), AND SYCIDIALES (M-O). J-L (AFTER PECK AND
EYER, 1963a). LEFT (APICAL); CENTER (SIDE); AND RIGHT (BASAL).

Genus *Stomochara* Grambast, 1961
  Stomochara moreyi (Peck) Grambast, 1961
Family Stellatocharaceae Conkin and Conkin, new family
  Genus *Leonardosia* Sommer, 1954
   Leonardosia langei Sommer, 1954

KEY TO ORDERS OF CHAROPHYTES

Enveloping cells vertical
  Unsegmented...Chovanellales
  Segmented, composed of polygonal plates................Sycidiales
Enveloping cells, spirally turned
  Dextrally..Trochiliscales
  Sinistrally...Charales

KEY TO GENERA OF NORTH AMERICAN PALEOZOIC CHAROPHYTES

Enveloping cells vertical
  Unsegmented..*Chovanella*
  Segmented, in polygonal patterns.........................*Sycidium*
Enveloping cells, spiral, and dextrally turned
  Five to 11 in number..................................*Moellerina*
Enveloping cells, spiral and sinistrally turned
  Eight to 13 in number, with large apical pore.............*Eochara*
  Six in number, with small apical pore.................*Palaeochara*
  Five in number with moderate to large apical pore......*Stomochara*

KEY TO NORTH AMERICAN SPECIES OF PALEOZOIC CHAROPHYTES

Vertical enveloping cells
  Unsegmented.................................*Chovanella burgessi*
  Segmented, composed of polygonal plates
   Sixteen enveloping cells....................*Sycidium clathratum*
   Eighteen enveloping cells...........................*S. foveatum*
Dextrally turned, spiral enveloping cells
  Nine to 11 enveloping cells; moderate development of cage
  superstructure; small (390 μ to 550 μ).......*Moellerina convoluta*
  Seven to 10 cells
   Moderately well developed cage superstructure;
   moderately large (552 μ to 1100 μ)...................*M. greenei*
   Well developed cage superstructure;
   subspherical; medium size (440 μ to 825 μ).........*M. bilineata*
   Excellently developed cage superstructure;
   subovoidal; large (750 μ to 1200 μ).............*M. laticostata*
Sinistrally turned, spiral enveloping cells
  Eight to 13 in number; large apical pore.......*Eochara wickendeni*

Six in number; small apical pore..............*Palaeochara acadica*
Five in number; moderately large apical pore....*Stomochara moreyi*

DISCUSSION OF CLASSIFICATION

Emendations

The class Charophyta is enlarged to include an additional and
new order, Chovanellales, which itself includes only the single
family Chovanellaceae Grambast, 1962 and the single genus *Chovanella* Reitlinger and Yartseva, 1958 (with *Chovanella kovalevi* Reitlinger and Yartseva, 1958 as the type species; from the Devonian of the U.S.S.R.). The new order *Chovanellales* is separated
from the other three orders of charophytes by its possession of vertical and unsegmented enveloping cells. Further, the gyrogonites
of *Chovanella* have been inverted by all workers except Conkin,
et al. (1974) who informally noted that the supposed apical coronula is, in actuality, the cage superstructure.

Grambast (1962) divided the family Porocharaceae Grambast, 1962
into two subfamilies: the Porocharoideae Grambast, 1961 (without a
utricle or an apical beak) and the Stellatocharoideae Grambast, 1962
(without a utricle, but with an apical beak). The apical beak is
an elongation of the apical portion of the enveloping cells; this
feature was fully discussed by Horn af Rantzien (1954, pp. 25-33)
at which time he placed *Stellatochara* within the Clavatoraceae, for
the apical beak is characteristic of the clavatoraceids. Grambast
(1962, p. 65) preferred to include the gyrogonites (with apical
beak, but no utricle) in the subfamily Stellatocharoideae. Inasmuch as *Palaeochara* Bell, 1922 possesses an apical beak (according
to Bell, 1922, p. 160 and Peck and Eyer, 1963b, p. 842), we prefer
to place *Stellatochara* Horn af Rantzien, 1954, *Leonardosia* Sommer,
1954, and *Masilovichara* Saidakovsky, 1962, together, and raise the
subfamily Stellatocharoideae Grambast, 1962 to the rank of a new
family Stellatocharaceae. We suggest that the stellatocharaceids
are derived from Palaeocharaceae Pia, 1927. We thus discern an
evolutionary lineage from Palaeocharaceae to Stellatocharaceae to
Clavatoraceae to Lagynophoraceae.

Inasmuch as *Stomochara* Grambast, 1961 has priority over *Catillochara* Peck and Eyer, 1963 (both names based on the concept of *Gyrogonites moreyi* Peck, 1934 and *Aclistochara moreyi* (Peck) Peck, 1937)
we must consider *Catillochara* to be a junior synonym of *Stomochara*
Grambast, 1961.

Catillochara Peck and Eyer, 1963 was erected to replace *Aclistochara moreyi* (Peck) Peck, 1937, and *Porochara moreyi* (Peck) Mädler,

1955; these gyrogonites have been consistently inverted by all previous workers; for example, Peck and Eyer's (1963b) illustrations of *Catillochara moreyi* are upside down (pl. 100, figs. 3-7; pl. 101, figs. 1, 6-8), except one figure (pl. 100, fig. 8). The genus *Stomochara* Grambast, 1961 (with corrected orientation) is used to include all post-Mississippian to pre-Cretaceous gyrogonites which are essentially of the charoideid kind except they possess a conspicuous apical opening which modern chareids do not. *Stomochara* belongs in the family Porocharaceae Grambast, 1962, which includes the following genera as well: *Praechara* Horn af Rantzien, 1954, *Porochara* Mädler, 1955, *Latochara* Mädler, 1955, and *Cuneatochara* Saidakovsky, 1962. Tentatively then, we accept the orientation of *Stellatochara* Horn af Rantzien, 1954, but believe that Peck (1934) and Peck and Eyer (1963b) have inverted the gyrogonite of *Stomochara* (*Catillochara* Peck and Eyer, 1963).

EVOLUTIONARY TRENDS

Peck (1936) discussed some evolutionary trends in the charophytes; the following trends are discerned presently.
1. Reduction of size of apical opening, beginning in the Jurassic, in the characeids and nitelloideids.
2. Reduction in number of enveloping cells, stabilized in the Pennsylvanian.
3. Very early trend to sinistral spiral enveloping cells, completed in the Pennsylvanian.
4. Increase in prominence of cage superstructure in the genus *Moellerina*, from Early Devonian through Early Mississippian (Kinderhookian).
5. Development of apical operculum (rosette) and surrounding dihescent suture in the dead end lineage, the raskyellaceids (Late Cretaceous-Late Oligocene).
6. Tendency for increased ornamentation until a utricle is developed in the clavatoraceids (Jurassic-Cretaceous) and lagynophoraceids (Upper Cretaceous-Paleocene); gyrogonites of clavatoraceids (and probably the lagynophoraceids) are like the stellatocharaceids, but unlike the porocharaceids.
7. Tendency for increased nodose ornamentation of gyrogonites in gyrogonids (Paleocene-Oligocene).
8. Trend to complexity of basal plates.
9. Double tiers of crown cells developed (probably in the Jurassic) as one half (Nitelloideae) of the dichotomous splitting of the porocharaceid line; the other line being the Charoideae with a single tier of crown cells.
10. Elongation of apical portions of the enveloping cells to form an "apical beak" in the stellatocharaceids; the apical beak is present in the gyrogonites of the clavatoraceids and this apical structure continues through the lagynophoraceids. This trend persisted from the Triassic to the Late Paleocene.

EVOLUTIONARY TREND IN THE CAGE SUPERSTRUCTURE

    A particular evolutionary trend, stratigraphically useful, is
the increase in prominence of the cage superstructure within the
North American representatives of the Early Devonian-Early Mississippian *Moellerina* (Conkin, *et al*., 1974).   The cage superstructure
is only moderately well developed in *Moellerina greenei* (Text-figs.
2G-I and 3) from the Onondagan Jeffersonville Limestone (the Lower
Paraspirifer acuminatus Subzone of Conkin and Conkin's 1972 revised
P. acuminatus Zone) of Jefferson County, Kentucky and southern Indiana (Text-fig. 4).   A few specimens of *M. greenei* are present in
Conkin and Conkin's 1972 third bone bed of the Middle *P. acuminatus*
Subzone of the *P. acuminatus* Zone (Text-fig. 4).   *M. convoluta*
(Peck), 1936, from the Middle Devonian of Missouri, is closely re-
lated to *M. greenei*, and, as such, possesses only a moderately well

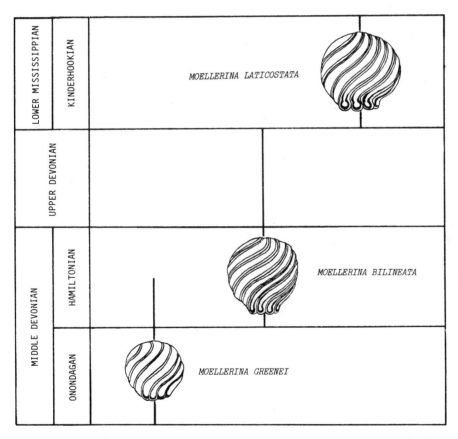

TEXT--FIGURE 3.   EVOLUTIONARY TREND IN THE CAGE SUPERSTRUCTURE
OF *MOELLERINA*.

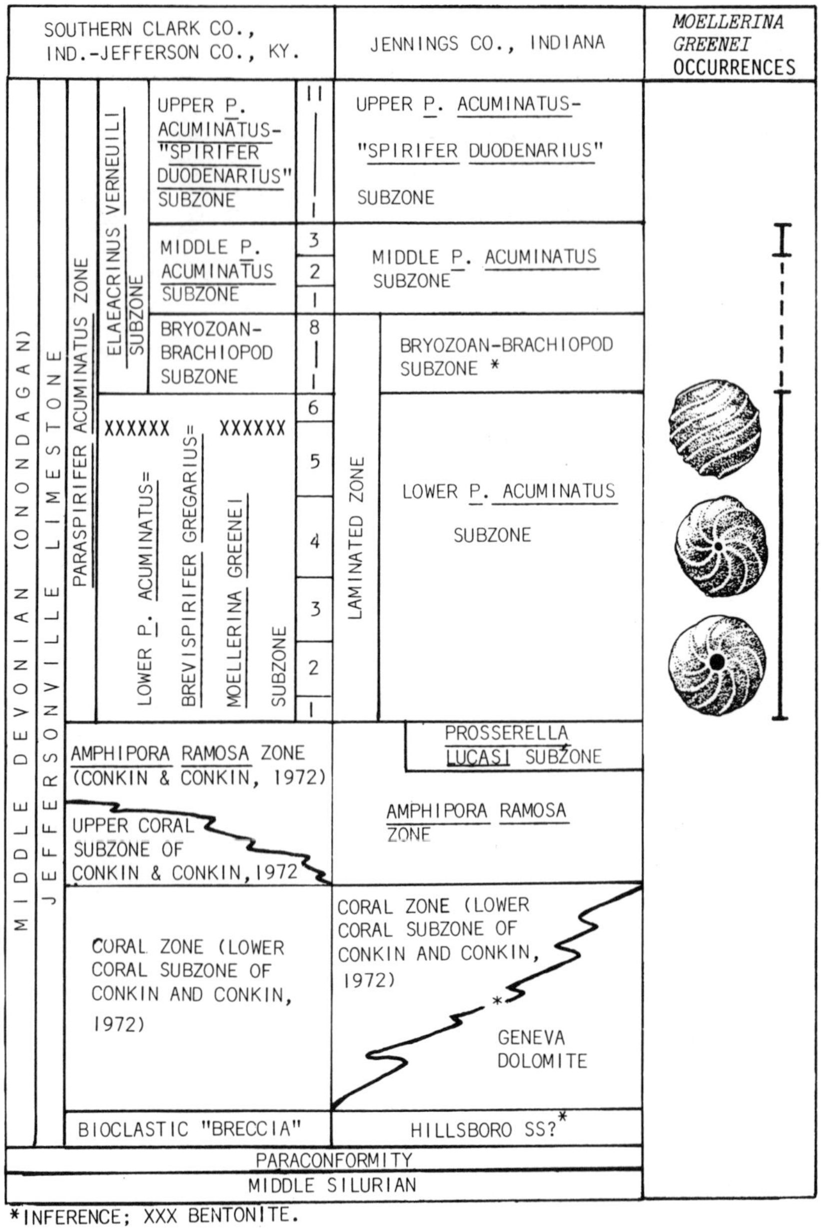

SOUTHERN CLARK CO., IND.-JEFFERSON CO., KY.					JENNINGS CO., INDIANA	MOELLERINA GREENEI OCCURRENCES

TEXT-FIGURE 4. STRATIGRAPHIC POSITION OF *MOELLERINA GREENEI* IN THE MIDDLE DEVONIAN (ONONDAGAN) OF NORTHWESTERN KENTUCKY AND SOUTHERN INDIANA (AFTER CONKIN AND CONKIN, 1975b).

developed cage superstructure as shown by Peck and Morales (1966, pp. 314, 315, pl. 2, figs. 7, 9). *M. convoluta* is in essential morphologic agreement with *M. greenei*; however, it is smaller (which by itself may not be a definitive specific character, particularly inasmuch as *M. convoluta* was based on an extremely small sample), but it does possess a larger number of spiral cells for its size (which is more significant) than does *M. greenei*. The cage superstructure is somewhat better developed in specimens of *M. greenei* from the lower Hamiltonian Dundee Limestone of Ohio (Conkin *et al.*, 1974, pl. 2, figs. 2, 3, 5, 6, 8, 9, 11, 12); these Dundee Limestone examples were reported by Kesling and Boneham (1966) as *Trochiliscus bellatulus* Peck, 1934. *T. bellatulus* was placed correctly in synonymy with *M. greenei* by Peck and Morales (1966, p. 315); however, the report of *M. bilineata* from the Dundee Limestone by Peck and Morales (1966, p. 313) is incorrect for this form also is referable to *M. greenei*. The cage superstructure characteristic of *M. greenei* is distinctly visible on Kesling and Boneham's figured specimens from the Dundee Limestone (1966, pl. 1, figs. 1, 4, 11, 21, 23, 24, 27-29, 31; pl. 2, figs. 1, 4, 7, 10, 13, 16, 19, 22, 24, 25, 28, 31, 34, 37, 40, 43, 46; pl. 3, figs. 1, 4, 7, 10, 13, 16, 19, 22, 25, 28, 31, 34, 37, 40, 43, 46). The cage superstructure is prominently developed in the late Middle Devonian (Hamiltonian) and the Late Devonian *M. bilineata* (Text-fig. 3) and excellently in the Early Mississippian (Kinderhookian) *M. laticostata* (Text-figs. 2D-F and 3).

PHYLOGENY

Difficulty has been experienced in attempting to erect a model of the evolutionary derivation of modern kinds of charophytes from the primitive forms. Particularly, the inverted orientation of Devonian, Early Mississippian, Pennsylvanian-Permian, and some Mesozoic gyrogonites made it difficult to correlate the features of charoideid and primitive charophytes. The apical and basal features of charophytes (both the oogonia of modern charophytes and the gyrogonites of fossil charophytes) are analyzed in Table 1 and illustrated in Text-figures 1-3, and 5.

As a result of the method of gyrogonite orientation presented by Conkin, *et al.* (1970, 1972, and 1974), *Karpinskya* Grambast, 1962 has been suppressed as a synonym of *Moellerina* Ulrich, 1886 by demonstrating that the "coronula" of *Karpinskya* is a cage superstructure (Text-figs. 2D-F) and also *Chovanella* Reitlinger and Yartseva, 1958 does not possess a preserved coronula, but that structure previously interpreted as a coronula is a cage superstructure (Text-fig. 2J-L). *Moellerina* Ulrich 1886, *Sycidium* Sandberger, 1849, and *Chovanella* Reitlinger and Yartseva, 1958 have been correctly oriented by Conkin, *et al.* (1970, 1972, and 1974) respectively and thus, no crown cells (coronula) are now known to be preserved

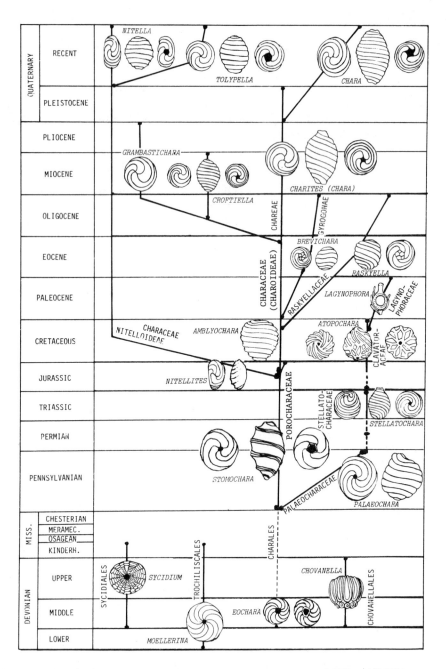

TEXT--FIGURE 5.   ILLUSTRATIONS OF SELECTED CHAROPHYTE GENERA
WITH PHYLOGENETIC INFERENCES.   THE VIEWS ARE GENERALLY ARRANGED
IN ORDER FROM LEFT TO RIGHT, APICAL, LATERAL, AND BASAL.

in Paleozoic charophytes. The recognition of the cage superstructure allows determination of the base of charophyte gyrogonites, which, in turn permits meaningful evolutionary studies of basal and apical structures by which the phylogeny of charophytes is simplified (Text-fig. 5) such that the ascent of charophytes from the primitive Eocharaceae is seen to proceed through the porocharaceids into the charoideids. The palaeocharaceids branched off from the porocharaceids some time in the interval Middle Devonian to Middle Pennsylvanian and the palaeocharaceids gave rise to the stellatocharaceids in the Permian (possibly through *Leonardosia* Sommers, 1954, a genus known only from South America) and these to the dead end evolutionary lineage, the clavatoraceids which branched off in Early Jurassic and in turn gave rise to the obscure lagynophoraceids (Upper Cretaceous-Paleocene). The operculated raskyellaceids of Late Cretaceous to Late Oligocene times are considered to be a dead end line derived from the charoideids (chareids). A fundamental dichotomy of the porocharaceids occurred in the Late Jurassic at which time the single-tiered coronula charophytes (the Charoideae) and the double-tiered coronula charophytes (the Nitelloideae) branched off and continued their independent evolutionary lines until the present day, becoming co-dominant in the Cenozoic. The chareid stock gave rise to the gyrogonids (Paleocene-Upper Oligocene) and two chareid groups: the *Croftiella-Grambastichara* kind (Upper Eocene-Pliocene), and the *Chara* kind probably by way of *Charites* (the Upper Eocene-Pliocene gyrogonites which are indistinguishable from the oogonia of *Chara*) which we tentatively consider to be, in actuality, congeneric with *Chara* (Quaternary).

If this scheme of orientation employed herein (Conkin, *et al.*, 1970, 1972, and 1974) is accepted as correct, then the phylogenetic and stratigraphic evidence is harmonious. Further, we can then suggest that the stellatocharaceids are not clavatoraceids, but that the clavatoraceids evolved from the stellatocharaceids, and, the lagynophoraceids were derived from the clavatoraceids (Text-fig. 5). Further, the stellatocharaceids may have been derived from the palaeocharaceids if *Palaeochara acadica* Bell, 1922 possesses an apical beak as Peck and Eyer (1963b, p. 842) believed. The Porocharaceae then can be derived from *Eochara* (Eocharaceae) by reduction of the number of enveloping cells to five and diminution of the apical opening. The porocharaceids are the stock from which all charoideid and nitelloideid charophytes evolved.

Stellatochara is restricted to the Triassic; the South American Permian stellatochareid genus, *Leonardosia* Sommers, 1954, may have given rise to the Jurassic-Cretaceous Clavatoraceae. *Stomochara moreyi* Grambast, 1961=*Catillochara moreyi* (Peck) Peck and Eyer, 1963 (1963b, pl. 100, figs. 3-8; p. 101, figs. 1, 6-8) and the form Peck and Eyer (1963b, pl. 101, figs. 2-5) referred to *Stellatochara prolata* Peck and Eyer, 1963 do not agree morphologically with

the gyrogonites of *Stellatochara* Horn af Rantzien, 1954 (1954, pl.
1, figs. 1-9; pl. 2, figs. 1-7, pl. 3, figs. 1-5, pl. 4, figs. 1-3);
this discrepancy is borne out in Horn af Rantzien's (1954 pl. 6,
figs. 3, 4) illustrations of Peck and Eyer's 1963 *Catillochara moreyi*=
Gyrogonites moreyi Peck, 1934. We consider *Catillochara* Peck and
Eyer, 1963 to be a junior synonym of *Stomochara* Grambast, 1961.

Further, as an indirect result of correcting Paleozoic charophyte gyrogonite orientation, it was discovered (Conkin, *et al.*,
1970) that *Weikkoella* Summerson, 1958 (a supposed foraminiferan) was
based on siliceous fillings of the oosporangia of the charophyte
Moellerina greenei Ulrich, 1886, emend. Conkin, *et al.*, 1970.

STRATIGRAPHIC RANGES OF PALEOZOIC PRIMITIVE CHAROPHYTES

In North America, primitive charophyte genera occur in the
Middle Devonian through Lower Mississippian (Kinderhookian); how-
ever, the oldest definite charophyte is in the presumed Downtonian
equivalent in Poland (Croft, 1952), which is considered to be Late
Silurian on the basis of ostracods and brachiopods, but which may be
Early Devonian. Actually a form of doubtful nomenclatural status,
Pseudosycidium Hacquaert, 1932, has been reported from the "Silur-
ian" of Sinkiang, but no real details are known on this find.

The first reference to charophytes in North America (Meek, 1871)
was to the subsequently named *Moellerina greenei* Ulrich, 1886 from
the Columbus Limestone of northern Ohio (actually the Lower *Paraspirifer acuminatus-"Spirifer macrothyris"* Zone; Text-fig. 6).

Oldest sinistrally spiralled charophyte is *Eochara wickendeni*
Choquette, 1956 from the Middle Devonian of Canada (Choquette, 1956).

Chovanella Reitlinger and Yartseva, 1958, represented by one
North American species, *C. burgessi*, is known from the Middle De-
vonian of North America; it ranges into the Upper Devonian of the
U.S.S.R. (Reitlinger and Yartseva, 1958).

Moellerina Ulrich, 1886, embracing the concepts of *Trochiliscus*
and *Eotrochiliscus* of Croft (1952) and *Trochiliscus* and *Karpinskya*
of Grambast (1962) is known from the North American Middle Devonian
(Onondagan) to Lower Mississippian (Kinderhookian), but *Moellerina*
is known in the Lower Devonian (or Upper Silurian?) of Poland (Croft,
1952). The type specimens of *M. greenei* are from Unit 5 (the silicified layer) of the Lower *Paraspirifer acuminatus* Subzone=(the
Brevispirifer gregarius Subzone=the *Moellerina greenei* Subzone) of
the *Paraspirifer acuminatus* Zone of the Jeffersonville Limestone
(Text-fig. 4) of Conkin and Conkin (1972, 1973b, 1974, and 1976).

Sycidium Sandberger, 1849, is known from the Late Devonian Lime Creek Formation of Iowa and the Early Mississippian of Missouri; it is present in the Lower Carboniferous of the U.S.S.R.; *Pseudosycidium* Hacquaert, 1932 may be a *Sycidium*. The only North American locality where representatives of *Sycidium* may be obtained at present is in the Lillibridge Quarry, near Mason City, Iowa (Conkin, *et al.*, 1972).

Palaeochara acadica Bell, 1922 is known only from the Pennsylvanian of Nova Scotia. *Palaeochara* differs from the Pennsylvanian porocharaceids and stellatocharaceids in having six sinistral spirals instead of five, but is like the stellatocharaceids in possessing an apical beak structure (according to Peck and Eyer, 1963, p. 482) and thus seems to be the derivative stock of the stellatocharaceids, clavatoraceids, and lagynophoraceids.

Stellatochara Horn af Rantzien, 1954 is known from the Triassic; this genus is differentiated from those members of the family Porocharaceae in that it possesses an "apical beak" which the porocharaceids do not. Even though some porocharaceids have been described as *Stellatochara* (such as *Stellatochara prolata* Peck and Eyer, 1963), we consider this to be incorrect and a result of inverted gyrogonite orientation. With this morphologic restriction, those porocharaceid forms which occur in the Pennsylvanian are excluded from *Stellatochara* Horn af Rantzien, 1954.

STRATIGRAPHIC ZONATION BY *MOELLERINA*

The genus *Moellerina* Ulrich, 1886, emend. Conkin, *et al.*, 1974, is present and can be used for zonation within the Middle and Upper Devonian and the Lower Mississippian of the United States (Text-fig. 3); four species are known in North America: *M. greenei* Ulrich, 1886 emend. Conkin, *et al.*, 1974 from the Middle Devonian of Indiana, Kentucky, Ohio, Wisconsin, and southern Ontario (Text-figs. 2G-I, 3, 4, and 6); *M. bilineata* (Peck), 1934, emend. Conkin and Conkin, 1974 from the upper Middle Devonian (Hamiltonian) Bell Shale of Michigan and the Upper Devonian of Missouri and Iowa (Text-fig. 3); and *M. laticostata* (Peck), 1934, emend. Conkin and Conkin, 1974 from the Lower Mississippian (Kinderhookian) Bachelor Sandstone of Callaway County, Missouri (Text-figs. 2D-F and 3). *M. convoluta* (Peck), 1936 from the Middle Devonian Callaway Limestone of Missouri may be a junior synonym of *M. greenei*, but presently we accept it as a separate species of *Moellerina*; it possesses a moderate development of the cage superstructure as it should, occurring as it does in the Middle Devonian.

The *Moellerina greenei* Zone (Conkin, *et al.*, 1970) is useful for long range correlation (Text-figs. 4 and 6) for it coincides with the Lower *Paraspirifer acuminatus* Subzone (=the *Brevispirifer*

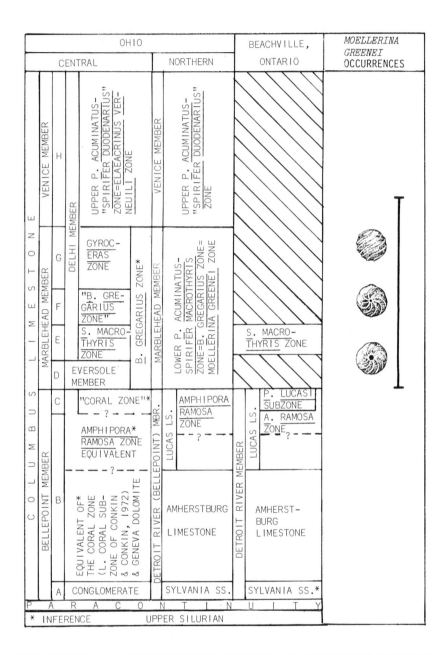

TEXT-FIGURE 6. STRATIGRAPHIC POSITION OF *MOELLERINA GREENEI*
IN CENTRAL AND NORTHERN OHIO AND SOUTHERN ONTARIO, CANADA
(AFTER CONKIN AND CONKIN, 1975b).

gregarius Subzone) of the Jeffersonville Limestone in northwestern
Kentucky and southern Indiana (Text-fig. 4), is correlative with the
D, E, F, and G zones of the Columbus Limestone of central Ohio (Conkin, *et al.*, 1970) and the Marblehead Member (Conkin and Conkin,
1975b) of the Columbus Limestone at Sandusky and Kelleys Island, Ohio,
and the E Zone of the Marblehead Member at Beachville, southwestern
Ontario (Text-fig. 6). Nevertheless, in rare numbers, *M. greenei*
does occur somewhat higher stratigraphically, in the H Zone of the
Columbus Limestone (Text-fig. 6) in central Ohio (Conkin, *et al.*,
1970) and in Unit 3 of the Middle *Paraspirifer acuminatus* Subzone
of the Jeffersonville (Text-fig. 4) in Kentucky (Conkin and Conkin,
1972). *M. greenei* does occur in the lowest Hamiltonian Dundee Limestone of Lucas County, northwestern Ohio and in the Hamiltonian
Milwaukee Formation at Milwaukee, Wisconsin, but the cage superstructure in these early Hamiltonian forms of *M. greenei* is somewhat more developed than in the Onondagan forms of *M. greenei*;
however, it is not as well developed as in *M. bilineata* of the
upper Hamiltonian. The record of *Moellerina bilineata* (Peck), 1934,
by Peck and Morales (1966, p. 313) in the Onondagan Jeffersonville
Limestone (actually from Conkin and Conkin's 1972 *Moellerina greenei*
Zone=*Brevispirifer gregarius* Subzone=Lower *Paraspirifer acuminatus*
Subzone of the *P. acuminatus* Zone) at the Falls of the Ohio, Jefferson County, Kentucky and southern Clark County, Indiana, in the
Onondagan Columbus Limestone (Stauffer's 1909 E-G Zones = Swartz's
1907 Marblehead Member of the Columbus Limestone) at Marblehead
Quarry, Sandusky, Ohio, and in the lower Hamiltonian Dundee Limestone, near Sylvania, Lucas County, Ohio, are in reality, references to *M. greenei* Ulrich, 1886; again, the cage superstructure
of these forms is not well developed as in *M. bilineata*.

LITERATURE CITED

Bell, W. A.   1922. A new genus of Characeae and new Merostomata
     from the Coal Measures of Nova Scotia. Roy. Soc. Canada Proc.
     and Trans. 16: 159-168.
Choquette, G. 1956. A new Devonian charophyte. Jour. Paleont.
     30: 1371-1374.
Conkin, J. E., and B. M. Conkin. 1972. Guide to the rocks and
     fossils of Jefferson County, Kentucky, southern Indiana, and
     adjacent areas. Univ. of Louisville Printing Service, Louis-
     ville, Kentucky.
_____, and _____. 1973a. The paracontinuity and the
     determination of the Devonian-Mississippian boundary in the
     type Lower Mississippian area of North America. Univ. of
     Louisville Stud. in Paleont. and Strat. No. 1. Univ. of Louis-
     ville Reprod. Serv. Louisville, Kentucky.
_____, and _____. 1973b. The *Amphipora ramosa*
     Zone and its significance in Middle Devonian stratigraphy of

east-central North America. Earth Science 1: 31-40.

Conkin, J. E., and B. M. Conkin. 1974. Supplement and Index to
Guide to the rocks and fossils of Jefferson County, Kentucky,
southern Indiana, and adjacent areas. Univ. of Louisville
Reprod. Serv. Louisville, Kentucky.

_____, and _____. 1975a. The Devonian-Mississippian
and Kinderhookian-Osagean boundaries in the east-central United
States are paracontinuities. Univ. of Louisville Stud. in
Paleont. and Strat. No. 4. Louisville, Kentucky.

_____, and _____. 1975b. Middle Devonian bone beds
and contact in central Ohio. Bull. Amer. Paleont. 67: 99-122.

_____, and _____. 1976. Guide to the Rocks and Fos-
sils of Jefferson County, Kentucky, southern Indiana, and adjacent areas, Second and Revised Edition. Univ. of Louisville
Reprod.Serv. Louisville, Kentucky.

_____, _____, G. S. Gregory, and A. L. Hotchkiss.
1974. Revision of the charophyte genus *Moellerina* Ulrich,
1886 and suppression of the genus *Karpinskya* (Croft, 1952)
Grambast, 1962. University of Louisville Stud. in Paleont.
and Strat. No. 3. Louisville, Kentucky.

_____, _____, T. Sawa, and J. M. Kern. 1970. Mid-
dle Devonian *Moellerina greenei* Zone and suppression of the
genus *Weikkoella* Summerson, 1958. Micropaleont. 16: 399-406.

_____, T. Sawa, T. C. Coy, and A. M. Salman. 1972. The
charophyte genus *Sycidium* in the Upper Devonian of Iowa. Micropaleont. 18: 75-80.

Croft, W. N. 1952. A new *Trochiliscus* (Charophyta) from the Downtonian of Podolia. Brit. Mus. (Nat. Hist.) Bull. 1: 180-222.

Grambast, L. 1962. Classification de l'embranchement des Charophytes. Naturalia Monspeliensia, Ser. Bot. 14: 63-86.

Harris, T. M. 1939. British Purbeck Charophyta. Brit. Mus. (Nat.
Hist.). London.

Horn af Rantzien, H. 1954. Middle Triassic Charophyta of South
Sweden. Opera Bot. 1:2: 5-83.

Kesling, R. V., and R. F. Boneham. 1966. *Trochiliscus bellatulus*
Peck from the Middle Devonian Limestone of northwestern Ohio.
Cont. Mus. Paleont., Univ. Mich. 20: 179-194.

Kidston, R., and W. H. Lang. 1921. On Old Red Sandstone plants,
showing structure, from the Rhynie Chert Bed, Aberdeenshire;
Part V, Roy. Soc. Edinburgh, Trans. 52: 855-906.

Meek, F. B. 1871. Descriptions of new species of invertebrate
fossils from the Carboniferous and Devonian rocks of Ohio.
Acad. Nat. Sci. Philadelphia, Proc. 23: 57-93.

Peck, R. E. 1934. The North American trochiliscids, Paleozoic
Charophytes. Jour. Paleont. 8: 83-119.

_____. 1936. Structural trends of the Trochiliscaceae. Jour.
Paleont. 10: 764-768.

_____. 1946. Fossil Charophyta. The American Midland Natural-
ist. 36: 275-278.

Peck, R. E.   1953.   Fossil charophytes.   The Bot. Review.   19(4):
    209–227.
_____, and J. A. Eyer.   1963a.   Representatives of *Chovanella*,
    a Devonian charophyte, in North America.   Micropaleont.   9:
    97–100.
_____, and _____.   1963b.   Pennsylvanina, Permian, and
    Triassic Charophyta of North America.   Jour. Paleont.   37:
    835–844.
_____, and G. A. Morales.   1966.   The Devonian and Lower Mis-
    sissippian charophytes of North America.   Micropaleont.   12:
    303–324.
Reitlinger, E. A., and M. V. Yartseva.   1958.   Nouye Kharofity
    verkhnefamenskikh otlozhenii Russkoi Platformy.   Acad. Nauk
    S. S. S. R., Doklady.   123: 113–116.
Stauffer, C. R.   1909.   The Middle Devonian of Ohio.   Geol. Surv.
    Ohio, Bull.   10.
Swartz, C. K.   1907.   The relations of the Columbus and Sandusky
    Formations of Ohio.   John Hopkins Univ. Circ., N. S. 7: 56–65.

CALCIFICATION OF FILAMENTS OF BORING AND CAVITY-DWELLING ALGAE, AND THE CONSTRUCTION OF MICRITE ENVELOPES

David R. Kobluk

McMaster University

Department Of Geology, Hamilton, Ontario, Canada

ABSTRACT

Endolithic (boring) algae play a significant role in the breakdown and alteration of carbonate skeletons, other carbonate structural elements, and sediment grains in reef environments. The activities of endolithic algae affect or control particle angularity and size, sediment porosity and permeability, particle micritization, micrite envelope formation, and other aspects of carbonate erosion and diagenesis.

In controlled experiments in the shallow marine environment at Discovery Bay, Jamaica, endolithic algae boring into crystals of Iceland spar calcite begin to grow out of the bores in the crystals into the sea after 25 days; after 65 to 95 days in the sea, dead and exposed filaments are completely calcified by the precipitation of micrite - size crystals of rhombohedral low Mg calcite on the exterior and interior of the thalli.

The coalescence of calcified algal filaments on the surface of grains will produce a "constructive" micrite envelope, which differs from the envelopes produced by the boring - infilling mechanism described by Bathurst, by forming entirely on the grain exterior. The process of constructive envelope formation can be geologically rapid, occurring within a few years or less. Micrite envelopes of this type are found on carbonate grains from the modern in Jamaica and other islands, and in sediments from the Devonian of western Canada, and the Ordovician of eastern Canada.

INTRODUCTION

Since about 1845 (Carpenter, 1845) endolithic (boring) algae
have received increasing attention as important bioerosive and
diagenetic agents in marine carbonate environments (endolithic
algae here refers to filamentous algae which actively bore into
carbonate substrates: they are distinguished from chasmolithic
algae living in cavities not of their own making, and epilithic
algae which live on the substrate surface; Fig. 1). The endolithic algae have a long history, spanning the Phanerozoic (Pia,
1937; Hessland, 1949; Gatrall and Golubic, 1970; Lukas, 1973;
Kobluk and Risk, 1974).

The boring algae (and fungi) produce a number of significant
geologic effects, including: 1) porosity and possibly permeability
modification (Kobluk, 1975b, 1976), 2) modification of carbonate
grain size and shape, 3) selective removal of parts of the carbonate component in mixtures of terrigenous clastic and carbonate
sediments (Perkins and Halsey, 1971), 4) limestone coast erosion

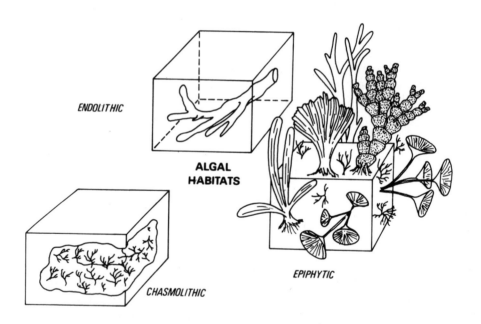

Fig.1. The three basic algal habitats. The endoliths actively
bore into the substrate; chasmoliths live in pre-existing cavities; epiliths (epiphytes) live on the surface of the substrate

(Purdy and Kornicker, 1958), 5) micritization of grains, the production of micrite envelopes on grains (Bathurst, 1964, 1966, 1971), and sparmicritization (Kahle, 1976), 6) probable effects on carbonate geochemistry by metabolite excretion. Because of their general dependency upon light, endolithic algae may also have value in paleoenvironmental interpretation in the delineation of paleobathymetry, and the depth of the paleophotic zone (Boekschoten, 1966; Swinchatt, 1969; Halsey, 1970; Perkins and Halsey, 1971; Colubic et al., 1975). However, the lack of dependance of endolithic fungi on light, and the great difficulty in distinguishing fungal from algal borings in ancient sediments and rocks, throw some doubt on the general applicability of this method.

One of the most significant and thoroughly studied endolithic algal phenomena is that of particle micritization and micrite envelope formation (micrite here refers to particles of less than 4um in size; the term originally referred to a microcrystalline fabric of coherent calcite crystals of less than about 4um or 5um in size; Folk, 1959). Bathurst (1966, 1971) described a process of repeated algal boring of carbonate grains, and infilling of the borings, presumably by precipitated micrite which resulted in a rind or envelope of micrite about the grains (Fig. 3), replacing the original grain periphery (see also, Winland, 1969; Friedman et al., 1971). Other micritization mechanisms have been proposed however, such as dissolution - repreciation in mucous coats (Kendall and Skipwith, 1969), and recrystallization (Purdy, 1968). The algal boring infilling mechanism noted by Bathurst (1966) is, however, common in shallow tropical marine carbonates. The carbonate infilling process is a function of carbonate saturation of the water. Endolithic algae are ubiquitous borers, but only in carbonate saturated or supersaturated waters is carbonate precipitated in the borings. In carbonate undersaturated waters the borings remain empty (Alexandersson, 1972), and micrite envelopes, and totally micritized grains, produced by algae, are therefore restricted to carbonate supersaturated waters (generally low latitude).

In attempting to gain a further understanding of micrite envelope formation and other endolithic algal effects, both experimental and observational approaches were used in this study. Carbonate substrates of known composition and structure were monitored with time to study the progressive development of endolithic algal infestation, and the various associated micritization phenomena. Modern and ancient reef carbonates were extensively sampled to provide naturally - occurring sample materials.

MATERIAL AND METHODS

Crystals of Iceland spar calcite were placed in the shallow
(60 cm) subtidal nearshore marine environment at Discovery Bay,
Jamaica (Fig. 2). Seventy - five crystals were harvested at intervals over a period of 257 days; the crystals were rinsed in dis-
tilled water, dried, and subsequently cleaned of organic matter in
a 15% solution of $CaCO_3$ buffered hydrogen peroxide. The cleaned
and dried crystals were studied under transmitted light, and coated
in aluminum for study on a AMR model 1000 scanning electron microscope. Compositional analyses were carried out using an EDAX
(energy dispersive x-ray analyser) unit on a Cambridge scanning
electron microscope and a KEVEX unit on a Hitachi scanning
electron microscope.

Holocene carbonate specimens were collected to a depth of 30 m
on the reefs at Discovery Bay Jamaica; at Malmok and Barcadera
Bonaire, Netherlands Antilles, Westpunt Baai Curacao, Netherlands
Antilles, Bellairs Barbados, Key Largo Florida, and North Point
and Whalebone Bay Bermuda. Pleistocene specimens were collected
from the Key Largo Limestone at Key Largo, the Miami Oolite at Big
Pine Key Florida, the first reef terrace at Rio Bueno and Discovery
Bay Jamaica, Westpunt Curacao, and Malmok Bonaire, and the 83,000

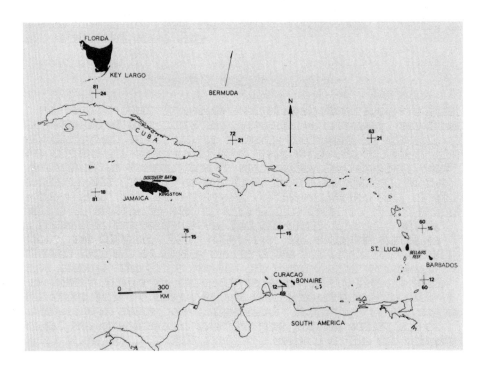

Fig.2. Location of study areas in the Caribbean

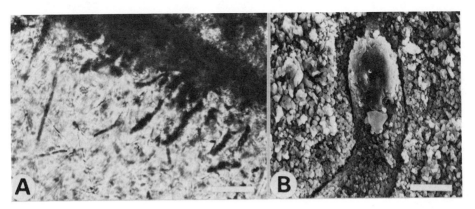

Fig.3. Micritization by the infilling of algal borings. (a) Micrite envelope and micrite-filled algal borings in a grain from the
Pleistocene of Barbados (83,000 year terrace). Scale bar 25um.
(b) Scanning electron micrograph of an algal boring filled by
centripetally-pricipitated micrite. Pleistocene of Barbados
(83,000 year terrace). Scale bar 4um

year terrace at River Bay and North Point, Barbados (Fig. 2).
Devonian specimens were collected from the Flume, Cairn, and
Southesk Formations of the Miette reef complex, Jasper, Alberta
(for a description, see Mountjoy, 1965; Noble, 1966; Kobluk,
1975a); Ordovician materials were collected from the Trenton
Formation, Montreal, Quebec, and from the Chazy Group at Isle
LaMotte, Vermont (for a description, see Kapp, 1975).

The Holocene, Pleistocene, Devonian, and Ordovician specimens
were coated in gold or aluminum and studied on a AMR model 1000
scanning electron microscope. Some of the Pleistocene, and all
the Devonian and Ordovician specimens were cut, polished, and
etched in 5% acetic acid prior to coating. All specimens were
studied in thin section and polished section; the Holocene and
Pleistocene materials were epoxy-impregnated prior to sectioning.

DESCRIPTION AND DISCUSSION

Endolithic algal infestation of the 75 Iceland spar crystals
was quantified over the 257 day period of the study, by point
counting unbored versus bored calcite in randomly - spaced parallel
transects on the spar crystals at 200 diameters (1000 counts on
each crystal). The results showed that endolithic algal infestation of the spar proceeds slowly with time for the first 15 days
after being placed in the sea, reaching only about 3% by surface
area at 15 days. Boring of the spar begins after about 7 days in

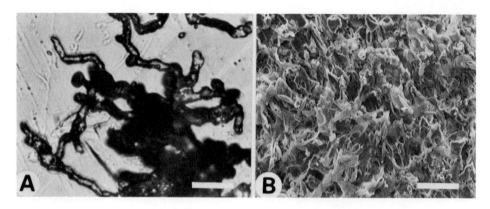

Fig.4. (a) Algal borings just below the surface in Iceland spar
calcite, following 31 days in the sea at Discovery Bay, Jamaica.
Scale bar 25um.   (b) The surface of Iceland spar calcite following
213 days in the sea in Jamaica.   Some of the borings contain precipitated micrite (not seen here).   About 30um of calcite has been
removed from the surface of the spar by the boring activity
(organic matter has been removed).   Scale bar 25um

the sea.   From 15 to 95 days, infestation proceeds rapidly, reaching 75% by surface area by day 95.   The rate of infestation subsequently levels off near the surface, and reaches 100% by 213
days (Fig. 4).

     After 25 days in the sea, endolithic algal filaments grow out
of their boreholes into the sea, approximately perpendicular to
the spar surface.   At 65 days, dead exposed endolithic algal
filaments are found completely encrusted, both internally and
externally, by rhombohedral micrite – size crystals of low Mg
calcite (Fig. 5).   Schroeder (1972) noted a similar growth out of
bores and calcification of filaments of the endolithic alga
Ostreobium quekettii in Bermuda reefs, but only within reef cavities.   The Bermuda filaments were therefore extensions of endolithic filaments living as chasmoliths, whereas those growing out
of the spar live as epiliths.

     The crystals filling in, and coating, the filaments on the
Iceland spar are all micrite – size.   Vacated algal borings within
the spar are also found partially filled (up to 25% by volume) by
precipitated micrite as early as 95 days;   they represent early
stages in the production of micrite tubules as described by
Bathurst (1966) in the boring – infilling micritization mechanisms (Fig. 2).

     Both the boring infilling and filament calcification are

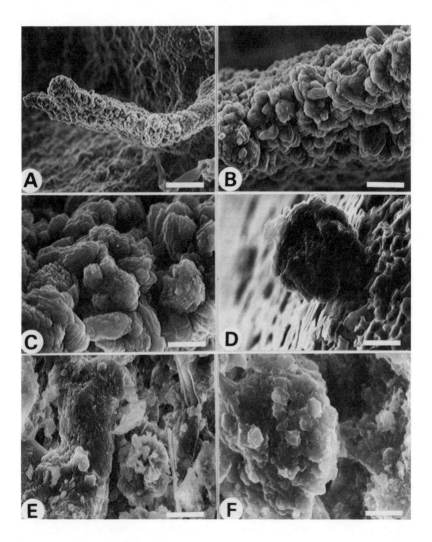

Fig.5. Calcified algal filaments. (a) Branching, completely calcified algal filament which had grown out of its borehole in
Iceland spar after 65 days in the sea in Jamaica. Scale bar 20um.
(b, c) Detail of the same calcified filament shown in (a) above,
showing the dense coating of low Mg micrite-size calcite crystals
Scale bars 4um and 2um respectively. (d) Short, calcified algal
stub, produced when a calcified filament has broken off, and
calcification continues. Scale bar 7um. (e) Two calcified filaments in an imcomplete constructive micrite envelope on a grain
from a depth of 20 m at Discovery Bay, Jamaica. Side view and
end-on view. Scale bar 4um. (f) Disc-like calcification developed
near the broken end of a calcified filament from a depth of 20 m
at Discovery Bay. Scale bar 2um

CONSTRUCTIVE   **DESTRUCTIVE**

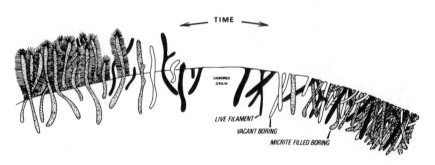

Fig.6. Comparison of constructive and destructive micrite envelopes. Destructive envelopes form by repeated algal boring of a
grain periphery, and infilling of the borings by precipitated
micrite. Constructive envelopes form by the coalescence of calcified algal filaments on the surface of the grain.

cementation phenomena, rather than processes associated with the
life activities of the algae. The infilled borings are always
found vacated, and the precipitation of carbonate occurs only on
dead filaments. To the author's knowledge, there are no studies
which have monitored the carbonate cementation process in a natural
environment, although there are many published estimates of the
rates at which cementation occurs, ranging from years to hundreds
of years (Martin and Ginsburg, 1965; Land, 1967; Purdy, 1968;
Macintyre et al., 1968, 1971; Shinn, 1969; Land and Goreau, 1970;
Bathurst, 1971; Schmalz, 1971). The current concensus is that
some cementation processes may occur within a few years (discussed
by Alexandersson, 1972). This study shows, however, that at least
in the case of some algae - associated cementation, weeks or months,
rather than years are required.

     In contrast to the boring - infilling formation of micrite
envelopes (Bathurst, 1964, 1966, 1971), micrite envelopes can form
outside, and on the surface, of a grain by the coalescence of
masses of calcified algal filaments such as those described above.
Algal filaments continue to grow through the grain surface to the
outside, and are calcified, until a sufficient filament density is
reached that the micrite coats on the filaments meet and form a
more or less continous envelope over the grain surface up to
several hundred microns thick (Fig. 6). The coalescence of

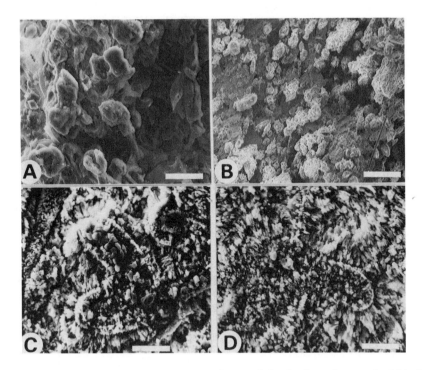

Fig.7. Scanning electron micrographs. (a) Coalescing calcified algal filaments on the surface of Iceland spar after 213 days in the sea at Discovery Bay, Jamaica. Scale bar 6um. (b) Calcified algal filaments on a carbonate grain from a depth of 20 m at Discovery Bay. The filaments show the development of calcified masses, an early stage in the formation of constructive micrite envelopes. Scale bar 40um. (c, d) Calcified algal filaments with-in constructive micrite envelopes on grains from the Pleistocene (83,000 year terrace) of Barbados. The interior fill, algal thallus, and palisade radial exterior cement are visible. Scale bars 7.5um and 7 um respectively

calcified filaments into complex masses is found to oocur on the Iceland spar as early as 213 days after placement in the sea (Fig. 7). These "constructive" envelopes are also found at various stages of development on carbonate grains from Jamaica, Curacao, Barbados, and Bonaire, at up to 20 m depth, and from the Pleistocene of Barbados and Key Largo, Florida (Fig. 7). Examples have also been found, of both calcified exposed endolithic algal filaments and constructive envelopes, from the Devonian of the Miette complex, and the Ordovician of Montreal and Isle LaMotte, Vermont.

Although the formation of constructive envelopes is a process occurring on the grain exterior, endolithic algae are of course involved, and there is no apparent reason why both constructive and "destructive" envelopes (the type described by Bathurst, 1966) could not develop simultaneously on the same grain. Indeed, examples are seen from the Pleistocene of Barbados, where this has occurred.

SUMMARY AND CONCLUSIONS

Iceland spar crystals monitored over a period of 257 days in the shallow subtidal marine environment at Discovery Bay, Jamaica, enable the processes of exposed endolithic algal filament calcification, algal boring infilling, and micrite envelope formation, to be observed taking place. Calcified algal filaments appear to be relatively common on shallow water carbonate grains from tropical marine environments. The endolithic algal filaments grow out of their bores, are calcified by low Mg calcite (on the low Mg Iceland spar) following death, and with time, reach a sufficient population density to form a coalescing mass of calcified filaments on the grain surface termed a "constructive" micrite envelope. Schroeder (1972) observed similar development in Bermuda reefs, but only on the walls of cavities, not on the exterior surfaces of grains.

The process of filament calcification and constructive envelope formation, like algal boring infilling, is a cementation phenomenon not related to the life processes of the algae. It serves to increase grain volume and surface area, reduce porosity, and possibly alter permeability, in sediments. Constructive envelopes will probably not form on grains which are constantly agitated. However, if the grains are stabilized by any of a number of possible mechanisms, such as algal scums, periodic quiet water conditions, or burial, for a few months to a few years, constructive micrite envelopes will have an opportunity to form.

ACKNOWLEDGEMENTS

The research, part of a Ph.D. thesis to be submitted at McMaster University, was supported by Dr. M.J. Risk, Department of Geology. Charles F. Kahle, Bowling Green State University, and Noel P. James, Memorial University, provided valuable comment throughout the study. The staffs of the Discovery Bay Marine Laboratory, Jamaica, the Caribbean Marine Biological Institute (CARMABI), Curacao, and the Bellairs Research Institute of McGill University, Barbados, and in particular Eileen Graham, Dr. Ingvar

Kristensen, and Dr. Finn Sander, were very helpful. Brian Pratt,
Bob Cavan, Michael Beattie, Elizabeth Kobluk, and Colin Yeo,
assisted in the field at various locations in the Rockies and the
Caribbean. Amoco Canada Petroleum Co. provided thin sections and
supported field work in the Rockies; Amoco Production Co., Tulsa,
provided thin sections. Dale Winland, Amoco Tulsa, and Michael
Rhodes, Amoco Canada, were particularly helpful during the initial
stages of the study. Financial support was available through Dr.
M.J. Risk, McMaster University, by grants from the National
Research Council of Canada, and the McMaster University Science
and Engineering Research Board. Financial support was also provided by the Department of Geology and Office of Graduate Studies,
McMaster University, and by Geological Society of America Penrose
research grants.

REFERENCES

Alexandersson, T., 1972, Micritization of carbonate particles:
    processes of precipitation and dissolution in modern shallow-
    marine sediments: Geol. Inst. Univ. Uppsala Bull., n.s.,
    v.3, p.201-236.
Bathurst, R.G.C., 1964, The replacement of aragonite by calcite in
    the molluscan shell wall: IN, J. Imbrie and N. Newell (eds.),
    Approaches to Paleoecology, New York, Wiley, p.357-376.
_____, 1966, Boring algae, micrite envelopes, and lith-
    ification of molluscan biosparites: Geol. Jour., v.5, p.15-32.
_____, 1971, Carbonate sediments and their diagenesis:
    Developments in Sedimentology 12, Amsterdam, Elsevier Pub.
    Co., 620p.
Boekschoten, G.J., 1966, Shell borings of sessile epibiontic organ-
    isms as paleoecological guides (with examples from the Dutch
    coast): Paleogeog. Paleoclimat. Paleoecol., v.2, p333-379.
Carpenter, W., 1845, On the microscopic structure of shells:
    British Assoc. Adv. Sci. Report, (1844), v.14, p.1-14.
Folk, R.L., 1959, Practical petrographic classification of lime-
    stones: Am. Assoc. Petroleum Geologists, v.43, p.1-38.
Friedman, G.M., Gebelein, C.D. and Sanders, J.E., 1971, Micrite
    envelopes of carbonate grains are not exclusively of photo-
    synthetic algal origin: Sedimentology, v.16, p.89-96.
Gatrall, M. and Golubic, S., 1970, Comparative study on some
    Jurassic and Recent endolithic fungi using a scanning electron
    microscope: IN, T.P. Crimes and J.C. Harper (eds.),
    Trace Fossils, Geol. Jour. Special Paper no.4, p.167-168.
Golubic, S., Perkins, R.D. and Lukas, K.J., 1975, Boring micro-
    organisms and microborings in carbonate substrates: IN,
    R.W. Frey (ed.), The Study of Trace Fossils, New York,
    Springer-Verlag Inc., p.229-260.

Halsey, S.D., 1970, Distribution and significance of marine boring endolithic algae and fungi in sediments of the North and South Carolina continental margin: M.A. thesis, Duke Univ., Durham, N.C., 85p.

Hessland, I., 1949, Investigation of the lower Ordovician of the Siljan district, Sweden.II: Lower Ordovician penetrative and enveloping algae from the Siljan district: Bull. Geol. Inst. Univ. Uppsala, v.33, p.409-424.

Kahle, C.F., 1976, Biogenic structures, carbonate cements, spar-micritization and evolution of Miami Limestone (Pleistocene) and calcareous crusts (Recent), Lower Florida Keys: Unpublished manuscript.

Kapp, U., 1975, Middle Ordovician stromatoporoid mounds in Vermont: Lethaia, v.8, p.195-207.

Kendall, C.G. St.C. and Skipwith, P.A. D'E., 1969, Holocene shallow water carbonate and evaporite sediments of Khor Al Bazam, Abu Dhabi, southwest Persian Gulf: Am. Assoc. Petroleum Geologists Bull., v.53, p.841-869.

Klement, K.W. and Toomey, D.F., 1967, Role of the blue-green alga Girvanella in skeletal grain destruction and lime mud production in the lower Ordovician of west Texas: Jour. Sedimentary Petrology, v.37, p.1045-1051.

Kobluk, D.R., 1975a, Stromatoporoid paleoecology of the southeast margin of the Miette carbonate complex, Jasper Park, Alberta: Bull. Canadian Petroleum Geol., v.23, p.224-277.

_____, 1975b, Endolithic algae in carbonates from the Caribbean (Abstract): Paleontology and Biostratigraphy Seminar, Toronto, Nov. 1975.

_____, 1976, Micrite envelope formation, grain binding, and porosity modification by endolithic (boring) algae in calcarenites in modern and ancient reef environments (Abstract): Geol. Assoc. Canada, Program with Abstracts, v.1, p.78.

_____, and Risk, M.J., 1974, Devonian boring algae or fungi associated with micrite tubules: Canadian Jour. Earth Sci., v.11, p.1606-1610.

Land, L.S., 1967, Diagenesis of skeletal carbonates: Jour. Sedimentary Petrology, v.37, p.914-930.

_____, and Goreau, T.F., 1970, Submarine lithification of Jamaican reefs: Jour. Sedimentary Petrology, v.40, p.457-462.

Lukas, K.J., 1973, Taxonomy and ecology of Recent endolithic microflora of reef corals with a review of the literature on endolithic microphytes: Ph.D. thesis, Univ. Rhode Island, 159p.

Macintyre, L.G., Mountjoy, E.W. and d'Anglejan, B.F., 1968, An occurrence of submarine cementation of carbonate sediments off the west coast of Barbados, W.I.: Jour. Sedimentary Petrology, v.38, p.660-663.

_____, 1971, Submarine cementation of carbonate sediments off the west coast of Barbados, W.I.: IN, O.P. Bricker (ed.), Carbonate Cements, The Johns Hopkins University Studies in Geology, v.19, p.91-94.

Martin, E.L. and Ginsburg, R.N., 1965, Radiocarbon ages of oolitic
    sands of Great Bahama Bank: Proc. Intern. Conf. Radiocarbon
    Tritium Dating, 6th Pullman, Wash., 1965, p.705-719.
Mountjoy, E.W., 1965, Stratigraphy of the Devonian Miette reef
    complex and associated strata, eastern Jasper National Park,
    Alberta: Geol. Survey Canada Bull. 110, 132p.
Noble, J.P.A., 1966, A plaeoecologic and paleontologic study of an
    Upper Devonian reef in the Miette area, Jasper National Park,
    Alberta, Canada: Ph.D. thesis, Western Reserve Univ.
Perkins, R.D. and Halsey, S.D., 1971, Geologic significance of
    microboring fungi and algae in Carolina shelf sediments: Jour.
    Sedimentary Petrology, v.41, p.843-853.
Pia, J., 1937, Die kalklosenden Thallophyten: Arch. Hydrobiol.,
    v.31, p.264-328; 341-398.
Purdy, E.G., 1968, Carbonate diagenesis: an environmental survey:
    Geologica Romana, v.8, p.183-228.
_____, and Kornicker, L.S., 1958, Algal disintegration of
    Bahamian limestone coasts: Jour. Geol. v.66, p.96-99.
Schmalz, R.F., 1971, Formation of beachrock at Eniwetok Atoll: IN,
    O.P. Bricker (ed.), Carbonate Cements, The Johns Hopkins
    University Studies in Geology, v.19, p.17-24.
Shinn, E.A., 1969, Submarine lithification of Holocene carbonate
    sediments in the Persian Gulf: Sedimentology, v.12, p.109-144.
Swinchatt, J.P., 1969, Algal boring: a possible depth indicator in
    carbonate rocks and sediments: Geol. Soc. America Bull., v.80,
    p.1391-1396.
Winland, H.D., 1968, The role of high - Mg calcite in the preserva-
    tion of micrite envelopes and textural features of aragonite
    sediments: Jour. Sedimentary Petrology, v.38, p.1320-1325.

AN AGRICULTURAL REVOLUTION IN THE LOWER GREAT LAKES

David M. Stothers   and   Richard A. Yarnell

University of Toledo   University of North Carolina

ABSTRACT

Until relatively recent times the earliest evidence of prehistoric agriculture in the Great Lakes region came from components
representing the proto-Iroquoian Princess Point Complex of southwestern Ontario. However, subsequent research has revealed earlier
evidence of plant husbandry by proto-Iroquoian peoples in northwestern Ohio and southeastern Michigan, who represent the late Middle
Woodland to early Late Woodland of the Western Basin Tradition.

Apparently, about 350-400 A.D. there occurred a climatic shift
as a result of a shift in atmospheric circulation patterns, which
brought about the end of the previous Sub-Atlantic period which
had been cooler and wetter or more severe. This atmospheric shift
brought about a climatic episode referred to as the Scandic episode,
which was a period of climatic amelioration. The climatic amelioration would have shifted the northern edge of the biotic province
within which maize agriculture had previously been practical, to an
even more northerly position, opening a new and hospitable 'frontier'.

The sociological implications of the settlement-subsistence
shift from seasonally wandering bands of hunters, gatherers, and
fishers to larger social aggregates of sedentary village agriculturalists are considered in terms of the social foundations for
later Iroquois culture.

The transformation of pre-agricultural levels of socio-cultural
integration to those attendant to sedentary agricultural village life,
eventually gave rise to socio-political sophistication such as

witnessed among the Huron, Neutral, and New York State Iroquois
Confederacies at the time of European contact. It would appear
that an agricultural revolution, as in other areas of the world,
was instrumental in attainment of even greater levels of socio-cultural sophistication, thus arguing in favour of the concept of cultural evolution.

This paper discusses this new research information with the
intent of generating theories to explain the source from which this
early agriculture, in the region of the Lake Erie drainage basin,
was derived; the possible reason for its introduction and adoption;
its impact on the already existing settlement-subsistence patterns;
and its concomitant ramifications on social organization.

                              INTRODUCTION

In this paper several radiocarbon dated archaeological sites
representing the time period between A.D. 500 and A.D. 1000, which
have revealed evidence of early maize husbandry will be considered.
These sites and others in the immediate area of concern have been
under archaeological investigation for the past seven years as part
of an ongoing research program into the derivation and advent of
early agricultural subsistence economy in the lower Great Lakes.

Carbonized remains of early agricultural produce, specifically
in the form of Zea mays or aboriginal Indian corn/maize, have been
recovered from three components of the early Late Woodland Princess
Point Complex in southwestern Ontario (Stothers 1973a, 1976), and
six components in southeastern Michigan (Stothers 1975e; Yarnell
1974, 1975a, b). Four of these latter sites pertain to the early
Late Woodland, Riviere aux Vase phase of the Western Basin Tradition (Stothers 1975a; Fitting 1965), while one site represents a
cultural manifestation called the Western Basin Complex (Stothers
and Prahl 1972; Stothers 1976, 1975a, b; Brose and Essenpreis 1973).
The Western Basin Complex is apparently a more lacustrine and littoral adapted manifestation which co-existed and interacted with
the Riviere aux Vase phase manifestation of southeastern Michigan
and northwestern Ohio, as well as the Princess Point Complex of
southwestern Ontario. These three manifestations: Riviere aux
Vase, Western Basin Complex, and Princess Point Complex represent
contemporary and interacting early Late Woodland expressions of a
cultural cline from southeastern Michigan and northwestern Ohio to
southwestern Ontario. The last of the six components from southeastern Michigan represents the late Middle Woodland transition
period of the Western Basin Tradition (Stothers 1975a,c,d). As
such, this last and earliest of the sites under consideration represents the period of transition from the late Western Basin Middle Woodland (Stothers 1975d) to the early Late Woodland Riviere

aux Vase phase of the Younge sequence (Stothers 1975c; Fitting 1965).

Early Agricultural Remains

The earliest dated maize in the Great Lakes region comes from the Indian Island No.4 site. This site is located on Indian Island which is located just north of the state boundary line between Michigan and Ohio (c.f. Map No.1) in Maumee Bay, in the extreme southwestern end of Lake Erie. This site is located in an undisturbed, sparsely forested area of the island, and has revealed two cultural strata which are separated by an intervening stratum of sterile lacustrine sand. Both cultural strata have produced diagnostic ceramics, lithics, and pit features, as well as maize kernels and cob fragments (Yarnell 1974, 1975a).

The lowest cultural stratum at this site has so far disclosed two carbonized maize kernels, each found in excavation units which are horizontally separated by some thirty feet. The uppermost cultural stratum has disclosed two fragments of a carbonized maize cob as well as sixteen carbonized maize kernels. The maize kernels were within a refuse pit which contained abundant cultural material and one of the radiocarbon samples to be discussed later in this paper. The above mentioned cob fragments were not recovered from within the pit, but were about two feet away, within the cultural stratum.

On the same island, about ¼ mile south of this site, is located the Indian Island No.3 site. This is a single component site, also located in an undisturbed and lightly forested area. A total of seventy-one carbonized maize kernels and kernel fragments, as well as five carbonized cob fragments were recovered (c.f. Table No.1) dispersed over an area of five hundred square feet of excavated ground (Yarnell 1974, 1975a). These kernels were intermixed with abundant ceramic and lithic cultural material. In addition to these kernels and cob fragments, three carbonized maize kernels were recovered from a cultural refuse pit feature, which also contained diagnostic cultural material and a radiocarbon sample.

Finally, the last site on Indian Island to yield carbonized cultigens was the Indian Island No.5 site (Yarnell 1975a). This site is located several hundred feet north of the Indian Island No.4 site, on the highest elevation of land on the entire island. Because of this particular location on the island, the same location has been repeatedly utilized as a favorite camping spot in recent times by hunters and recreational campers. This has led to this particular locale becoming the sole clearing on the island, and the prehistoric cultural remains at this site have been severely disturbed by countless modern pits dug for camp fires and burying refuse. Nonetheless, limited test trenching recovered one kernel and three cob fragments (c.f. Table No.1) of the prehistoric Northern Flint variety of maize in association with Riviere aux Vase phase ceramics.

Two other early Late Woodland, Riviere aux Vase phase sites which have revealed early cultigens, are on Gard Island, located about ½ mile to the east of Indian Island.

On Gard Island, two components have been excavated which produced remains of prehistoric cultivated plants. These sites are designated Gard Island No.2 and No.3.

The Gard Island No.2 site (Lozanoff and Stothers 1975) represents both a habitation and burial area for early Riviere aux Vase phase people. A total of 37 individuals were recovered from a small cemetery area, the graves of which had apparently been dug into the site after occupation had been shifted some one hundred feet further south. Burial vessels and vessel fragments in the graves, as well as radiocarbon dates derived from human bone from the graves are in typological and temporal agreement with the other scattered habitation refuse such as ceramic vessel fragments, ceramic pipes, and lithic tools. These facts in conjunction with the total absence of cultural material from earlier or later times argues in favor of an earlier habitation area being later used by the same people as a cemetery area.

At this site a total of four carbonized maize kernels were recovered. One maize kernel was recovered from within each of the grave pits for burial Features 5 and 6. In addition, a small pit feature (designated Feature 6) located near one edge of the cemetery revealed two additional carbonized maize kernels.

The other site on Gard Island, Gard Island No.3, is located at the very northernmost end of the island. This site like Gard Island No.2, which is located about 500 feet south of Gard Island No.3, is sparsely forested and overgrown with abundant scrub brush. Also like Gard Island No.2, Gard Island No.3 was undisturbed, the cultural stratum being sealed some 2½ to 3.0 feet below the present surface.

At Gard Island No.3, four carbonized maize kernels were recovered from the general cultural stratum in Trench No.3, while pit feature No.1 in this same trench disclosed one carbonized wild grape and one carbonized _Polygonum_ seed. These latter two seeds were in association with abundant fish bone, and some diagnostic ceramics. Two radiocarbon dates were determined from the wood charcoal sample in this pit, while one radiocarbon date was derived from scattered wood charcoal flecks recovered from the cultural stratum in Trench No.1.

The Sissung site, located further north of the previously discussed sites, along the western shore of Lake Erie, disclosed a few projectile points, and some diagnostic ceramics (Fitting 1970: 151), as well as carbonized hickory nut, carbonized maize and a wood charcoal radiocarbon sample (Fitting 1966, 1970; Yarnell 1964: 125, 195; Prahl 1969: 41-60). This early Late Woodland site was originally interpreted by Fitting as representing a cultural blend between the more western Wayne Tradition and the Late Woodland Riviere aux Vase phase, of what was then called the Younge Tradition (Fitting 1966, 1970: 151, 242). Subsequently, the Younge Tradition has been subsumed within a more inclusive and temporally extended taxon referred to as the Western Basin Tradition (Stothers 1975a). Nevertheless, the Sissung site has more recently been referred to as one of the components representing the early Late Woodland Western Basin Complex (Stothers and Prahl 1972; Stothers 1976; Brose and Essenpreis 1973: 74-75). This manifestation is apparently more lacustrine and littoral in its geographic distribution in southeastern Michigan and northwestern Ohio (Stothers 1975b: 120-121), and represents an intermediate cultural expression between the contemporaneous Princess Point Complex of southwestern Ontario (c.f. Map No.1), and the Riviere aux Vase expression of southeastern Michigan and northwestern Ohio.

The fact is that too little cultural material was recovered from the Sissung site to validate either the viewpoint of Fitting or Stothers, and this site may in effect represent a cultural overlap and admixture of the Wayne Tradition, the Riviere aux Vase phase, and the Western Basin Complex. Such a situation would not be unexpected in the western Lake Erie region, where it is known that all of these cultural expressions and many more were overlapping and interacting (Stothers 1973b, 1975e) within an area which served as a major cross-roads between areas to the east and west, as well as to the north and south.

North of the Lake Erie shore in southwestern Ontario, the early Late Woodland expression referred to as the Princess Point Complex (Stothers 1972, 1973a, 1976, 1975b) has disclosed evidence of carbonized maize in each of its three temporally sequential phases which span the time period between A.D. 600 and A.D. 950-1000 (Stothers 1976: 78-114).

Table No.1

Early Agricultural Remains Recovered from
Sites in the Lake Erie Drainage Basin

Agricultural Sites and Cultural Affiliation	Kernels	Cobs	Cob Fragments
TRANSITIONAL MIDDLE TO LATE WOODLAND IN THE WESTERN BASIN TRADITION			
Indian Island No.4			
Upper Stratum (2a)	16	-	2
Lower Stratum (3b)	2	-	-
LATE WOODLAND, WESTERN BASIN COMPLEX			
Sissung	2(?)		
LATE WOODLAND, RIVIERE AUX VASE PHASE OF WESTERN BASIN TRADITION			
Indian Island No.5	1	-	3
Indian Island No.3	71	-	5
Gard Island No.2	4	-	-
Gard Island No.3	4	-	-
LATE WOODLAND, PRINCESS POINT COMPLEX			
Porteous	44	1	-
Grand Banks	1	-	-
Princess Point	4 or 5	-	-

Radiocarbon Dating Early Agricultural Sites

The Indian Island No.4 site (previously discussed) which represents the period of transition between the late Western Basin Middle
Woodland, and the early Late Woodland Riviere aux Vase phase has produced four radiocarbon dates - three from the lowest (and earliest)
cultural stratum and one from the uppermost cultural stratum (Stothers
1975e).

Pit No.6, in the uppermost cultural stratum at this site contained the sixteen (16) carbonized maize kernels and two (2) nearby
and associated carbonized cob fragments, as already discussed. The
radiocarbon date obtained on wood charcoal from this pit was A.D.

540 + 95 (DIC-414) (Stothers 1975e).

The lowest cultural stratum at this site which produced the remaining three radiocarbon dates, disclosed the following information.
Cultural Pit No.1 produced one (1) carbonized maize kernel, diagnostic cultural material and a wood charcoal sample which rendered a
radiocarbon date of A.D. 840 + 85 (DIC-301). Pit No.2, located only
a few feet away in this lower cultural stratum, disclosed wood charcoal which produced a date of A.D. 650 + 75 (DIC-300). This charcoal
sample was also in association with diagnostic cultural material.

The remaining maize kernel recovered from this lowest stratum
was retrieved some thirty feet distant from the excavation unit which
produced radiocarbon dates DIC-301 and DIC-300. This single carbonized maize kernel was in association with diagnostic cultural material and scattered wood charcoal flecks. These charcoal flecks produced a carbon date of A.D. 590 + 55 (DIC-415).

All four of the carbon samples from this site were severely contaminated with rootlets, and samples DIC-300 and DIC-301 were cleaned
by hand-picking under high power magnification, while DIC-414 and
DIC-415 were not only hand-picked under high power magnification, but
they were also processed with a new chemical rootlet treatment. As
such the dates for DIC-300 and DIC-301, which are not consistent with
each other, are undoubtedly too recent because of rootlet contamination. The remaining two dates for DIC-414 and DIC-415 are entirely
acceptable.

Although DIC-414, which is from the upper cultural layer, dates
slightly older than DIC-415 from the lower cultural stratum, the cultural material from both cultural strata at this site are so similar,
that the real temporal difference between the two strata is, in all
probability, somewhat concealed by the standard range of error assigned to these two radiocarbon dates. As such, the lowest stratum
can comfortably be assigned to the earlier half of the sixth century
A.D., while the upper stratum can comfortably be assigned to the later half of the sixth century A.D.

Due to the extremely disturbed nature of the Indian Island No.5
site, no charcoal samples were processed to radiocarbon date the site.
However, cultural material would indicate a temporal placement of
A.D. 600-900 for this site.

The Indian Island No.3 site, located to the south of Indian Island No.4 (c.f. Map No.1), disclosed seventy-one (71) carbonized
maize kernels as well as five (5) carbonized cob fragments (Yarnell
1974; 1975a, b; Stothers 1975e) (see previous discussion). Three of
these maize kernels were recovered within a cultural refuse pit, in
association with ceramics and wood charcoal. The charcoal sample

produced a radiocarbon date of A.D. 960 ± 80 (DIC-413). This date
is entirely acceptable in view of the associated cultural material
in the pit and elsewhere at this single component site.

On Gard Island, to the east of Indian Island, two additional
single component Riviere aux Vase phase sites have produced radiocarbon dates.

One of these sites, Gard Island No.2, which revealed the four
(4) maize kernels, one in each of two grave features, and the remaining two kernels in the small associated pit feature, has produced two radiocarbon dates. A third sample is still being processed
and will be reported upon at a later date. All three of these dates
are derived from human bone, due to a paucity of contextually reliable wood charcoal.

The first of these dates, derived from human bone in Burial Feature No.4, was A.D. 610 ± 80 (DIC-416). In view of the typical Riviere aux Vase ceramics (Fitting 1965) found in abundance at this
site, this date is perfectly acceptable and agrees with ceramic seri-
ational estimates.

The second radiocarbon date received from this site was obtained
from human bone in Burial Feature No.3a. This date was A.D. 0 ± 70
(DIC-417). In view of the associated cultural material, this date
is much too early and obviously is not in close accord with the other
date (DIC-416). This second date is rejected as erroneous, although
a reasonable explanation for this date can not now be advanced, unless it is due to fossil carbonate contamination.

The remaining component on Gard Island, which has produced maize
and radiocarbon dates, is the Gard Island No.3 site. At this site
the two radiocarbon dates derived from wood charcoal in the cultural
pit which contained the carbonized wild grape and Polygonum seeds,
fish bone and diagnostic ceramics, were both from the same charcoal
sample, which was simply divided in half. This charcoal sample was
so severely contaminated by rootlet infestation that one half of the
sample was cleaned by hand-picking and sorting under high power magnification, while the other half of the sample was not only hand-·
picked and sorted, but it was processed with a new chemical rootlet
extraction procedure. The first sample, which had no chemical rootlet extraction undertaken on it, produced a date of A.D. 990 ± 75
(DIC-412b). The second half of the sample, which was chemically
treated, produced a date of A.D. 920 ± 90 (DIC-412a).

The third sample, from Trench No.1, rendered a date of A.D. 1090
± 50 (DIC-411).

In view of the very similar cultural material and temporally equivalent radiocarbon dates from Gard Island No.3 and Indian Island No.3, these two sites can best be assigned to the late tenth century, or terminal aspect of the Riviere aux Vase phase of the Late Woodland time period (Stothers, Miller and Berres n.d.).

The Sissung site, located further north along the western Lake Erie littoral, (c.f. previous discussion) has been dated to A.D. 700 ± 120 (M-1519).

North of the Lake Erie shore in southwestern Ontario, the early Late Woodland cultural manifestation referred to as the Princess Point Complex has disclosed evidence of early maize husbandry (Stothers 1973a).

The Princess Point Complex was represented, by three regionally distinct variations or 'foci' (Stothers 1972, 1976, 1975b). These foci have been called the Point Pelee, Grand River, and Ausable foci.

Although several radiocarbon dates have been received for components of the Point Pelee and Grand River foci, which bracket the Princess Point Complex between A.D. 600 and A.D. 950-1000 (Stothers 1976), agricultural remains have so far only been recovered from three components of the Grand River Focus. Each of these components represent one of the three temporally sequential phases within the temporal duration of this culture complex. The earliest phase is believed to have persisted from A.D. 600-750, the middle phase existed from A.D. 750-850, and the latest phase from A.D. 850-950. Thereafter, the Princess Point Complex is succeeded by the Glen Meyer culture (Wright 1966) which developed out of the earlier Princess Point base (Stothers 1976).

The Princess Point type site, located along the edge of a small peninsula, which projects into a marshy inlet at the western most end of Lake Ontario, (c.f. Map No.1) produced four (4) or five (5) charred maize kernels. This site is an early phase component of the Princess Point Complex.

The Grand Banks site, located on the edge of a floodplain along the Grand River about 20 miles upriver from its embouchure into Lake Erie, produced a single carbonized maize kernel (Cutler and Blake 1971) in a cultural pit. This site represents the middle phase of the Princess Point Complex.

The Porteous site represents the last or terminal phase of the Princess Point Complex. It was a small palisaded village site located on the summit of a small hill overlooking the Grand River about 30 miles upriver from its embouchure into Lake Erie. This site produced forty-four (44) carbonized maize kernels and a single 8-rowed carbonized maize cob (Stothers 1973a; Yarnell 1973). The kernels and

cob from the Porteous site were recovered from within house pits and
undisturbed midden deposits.

Early Maize Variety in SE Michigan and SW Ontario

The archaeological maize so far recovered from the previously
discussed components in southeastern Michigan and southwestern Ontario differs from the classic northern Flint/Eastern Complex race
(Yarnell 1973, 1974, 1975a, b; Cutler and Blake 1971, 1973).

Although, much of the previously mentioned maize displays similarities to the Northern Flint race, there are some apparent differences in the kernels and cobs.

Concerning the maize derived from the Indian Island No.4 site,
the earliest radiocarbon dated agricultural component in the Great
Lakes, Yarnell states (1975a) that these maize kernels are not of
the classic eight row Northern Flint variety, and suggests that
"Kernels appear to have changed in shape between Middle Woodland and
Late Woodland, and thereafter increased in size, especially in width,
while continuing the change in shape". In other words, the maize
from this Middle to Late Woodland transitional site probably represents an evolutionary stage of development in Northern Flint maize.
Earlier Middle Woodland varieties probably produced kernels of approximately equal length and width near the mid-ear, while Late Woodland varieties became wider, longer, and more 'kidney' shaped. Yarnell states (ibid), "It is my impression that cobs did not change
greatly in row number between Late Middle Woodland and early historic
times, but that cupules increased in size." These arguments are further supported by the data revealed by the maize kernels recovered
from the Gard Island sites as well as the kernels recovered from the
other Indian Island sites. These kernels are smaller than the later
'classic' Northern Flint, especially in width, indicating smaller
cobs, higher row number, or both (Yarnell 1975a).

Derivation of Early Great Lakes Maize

The author concurs with Wright (1972: 57) that "it was probably
through the Princess Point culture that corn agriculture was first
introduced to Ontario", and that the Princess Point culture "appears
to represent one of the late developments of the dramatically changing Hopewell cultures (Wright 1972: 53)".

The earlier Saugeen, Point Peninsula, Nottawasaga, and Western
Basin Middle Woodland cultural manifestations of southwestern Ontario,
southeastern Michigan and northwestern Ohio (Stothers 1975d) do not
appear to display cultural continuity with the later Princess Point

Complex. It has recently been suggested that the Princess Point
culture represents a cultural intrusion (Stothers 1973a, 1976, 1975b,
c) into southwestern Ontario from regions to the south. This cultural intrusion has been elaborated upon elsewhere in considerable
detail (ibid) and appears to represent a coeval and analogous situation to that of southeastern Ontario (Spence 1967; Johnston 1968a,
b), Pennsylvania (Jones 1931; McCann 1971), and New York State
(Ritchie 1938, 1969) where some aspect(s) of the fragmenting Hopewellian Interaction Sphere were "fusing with local resident complexes
after shifting into these other regions (Ritchie 1969: 215)".

The reasons for this colonial occupation of southwestern Ontario
by the Princess Point Complex people, and the corresponding appearance of the Western Basin Complex in the littoral regions of southeastern Michigan and northwestern Ohio will be considered in the
following section of this paper.

Despite the fact that maize agriculture was probably introduced
into Ontario during the period of colonization by the Princess Point
and Western Basin Complex peoples, the Riviere aux Vase and late
Western Basin Middle Woodland peoples of southeastern Michigan and
northwestern Ohio probably had already commenced limited maize husbandry as a result of the diffusion of the idea or concept of an
agricultural life-way from Hopewellian culture(s) to the south.
Such a contention is strongly supported by the maize recovered from
the Indian and Gard Island sites.

The Northward Spread and Adoption
of an Agricultural Lifeway

Apparently, about A.D. 350-400 there occurred a climatic shift
(Bryson and Wendland 1967; Baerreis and Bryson 1965) as a result of
a shift in atmospheric circulation patterns, which had been cooler
and wetter or "more severe" (Baerreis and Bryson 1965: 214; Bryson
and Wendland 1967: 294). This atmospheric shift brought about a
climatic episode referred to as the Scandic Episode (ibid: 215 and
294) which was a period of climatic amelioration.

The authors believe that the initial climatic amelioration of
the Scandic Episode may have made it possible for many of the cultural units which constituted the Hopewellian Interaction Sphere to
spread further afield, in search of regions in which they could pursue their previous lifeways, which included the initial experimental growth of maize. We know that some peoples of the Hopewellian
Interaction Sphere had maize (Vickery 1970; Struever and Vickery
1973; Prufer 1964a, b, c; Struever 1964), although some locales and
cultures within the Hopewellian Interaction Sphere may have been
more committed to a pattern of "Intensive Harvest Collecting" as

suggested by Struever (1968). Perhaps certain local environmental
adaptations favored intensive harvest collecting over maize husbandry, and possibly the cooler, wetter climatic conditions of the
earlier Sub-Atlantic period wouldn't allow for larger scale, universal adoption of maize agriculture, as is strongly implied by Yarnell
(1964: 149). If these Hopewellian cultures were aware of maize
husbandry, as evidence strongly indicates, and if they only practiced it to a limited degree in certain environmental locales
which offered a more advantageous adaptation, because of a restrictive climatic regime, the following Scandic Episode, which represents climatic amelioration, may have allowed Hopewellian peoples
to more actively pursue an experimental-developmental agricultural
way of life. The climatic amelioration would have shifted the northern edge of the biotic province within which maize husbandry had
previously been possible to an even more northerly position. This
sequence of events would have opened a new and hospitable 'frontier'
into which some of these people could have moved to 'colonize' and
actively pursue their lifeways. This is certainly what is suggested
by the climatic data available, as well as the roughly contemporaneous movement of Hopewellian derived cultures into Pennsylvania,
New York State, southeastern Ontario, and southwestern Ontario.
Because there is no apparent cultural continuity between the early
Late Woodland Princess Point Complex of southwestern Ontario and
the earlier Middle Woodland cultures (Stothers 1976, 1975c, d),
such an idea would readily explain the sudden and dramatic appearance of this 'full-blown' agricultural involvement to the north of
Lake Erie. Since the late Middle Woodland and Late Woodland peoples of southeastern Michigan and northwestern Ohio do not appear
to be a derivative from elsewhere (with the exception of the Ontario
related Western Basin Complex), but are a continuous, evolutionary
'in situ' development in the Western Lake Erie Basin (Stothers 1975a),
it would appear that maize husbandry was adopted by these people
as a result of the diffusion and adoption of the idea from regions
to the south.

The Demise of the Hopewellian Interaction Sphere

An adjunct paradigm of the preceding discussion of climatic
shifts and population dispersals, involves the correlated reasons
for the decline of the Hopewellian Interaction Sphere (Stothers
1976).

Perhaps the regional clustering of burial mounds into mound
complexes, in disparate, restricted areas, reflects more than the
lavish disposal of exotic goods by members of a sophisticated mortuary cult. Perhaps most of these exotic grave goods and the mound
complexes themselves, are reflections of the non-perishable aspects
of a series of trade and exchange systems, strengthened and enforced
by religious cult ideas, which developed as inter-group reciprocity,

in response to very demanding economic needs. These economic needs
may have been the result of a combination of a uniform and restrictive ecological environment in certain locales, as well as a restrictive cultural adaptation to the particular ecological parameters.
In other words, perhaps certain ecological regions were so uniform
and restrictive that many pursuits and activities, as well as potential natural resources were not available. Furthermore, some local
groups may not have cognitively adjusted to their ecological parameters in a manner that would relay a maximum degree of 'positive
feedback'.

Such a situation would depend on outside economic aid and support, in the form of trade and exchange, which in turn would require
inter-group reciprocity over expanse areas. In support of such a
contention is the fact that many of the regions that contain burial
mound complexes, or numerous individual burial mounds, and numerous
exotic mortuary items are ecologically restrictive (i.e. uniform/
non-diversified), while many interdispersed or surrounding regions
which are ecologically diverse, display an absence or near absence
of burial mounds and exotic trade goods. These regions of ecological diversity would potentially be more self-supporting (due to
widely varied subsistence resources) and would not need to rely on
outside resources (Brose and Essenpreis 1973: 73-74; Taggart 1967).

As such, climatic amelioration about A.D. 350-400, in conjunction with new and better adapted genetic strains of maize, which
probably slowly evolved in the northern environs of the American
Midwest, may have produced a much more reliable subsistence basis.
This more reliable and more productive subsistence basis, in conjunction with a northward shift of the northern edge of the biotic province in which prehistoric maize economy was possible may have aided
in the erosion of institutions which, in fact, constituted the Hopewellian Interaction Sphere. Thus, the development of a new and reliable subsistence basis would have rendered the old trade and exchange system of interdependent reciprocity groups impractical,
while the movement of some of these groups into the newly available
'frontier' lands would have further weakened the social, religious,
and economic ties which had previously existed. Only future research
will give greater insight into this problem.

This theory is in direct contradiction to the hypothesis of
James B. Griffin (1960: 21-22) who suggested:

> ". . . that a minor phase of cooler climate
> in the northern Mississippi Valley and
> Great Lakes areas was a significant contri-
> buting factor to a decline in the reliabil-
> ity of agricultural products, which in turn
> resulted in the decline from Hopewell, or

Hopewell-Woodland, to early Late Woodland
cultural forms."

It is believed that as peoples actually moved into these newly
available frontier lands, which were more distant from the northern
Mississippi Valley and its adjacent regions, the maintenance of earlier forms of social organization, and inter-dependent religious and
trade sphere institutions became increasingly difficult to maintain.
Coincidentally the growing use of, and reliance upon maize husbandry
rendered former ties of inter-dependent reciprocity meaningless, and
as a result the Hopewellian Interaction Sphere disintigrated.

Settlement-Subsistence Systems

In order to understand the settlement-subsistence systems of
these early agricultural peoples in southwest Ontario, southeast
Michigan and northwest Ohio the subsistence system must be viewed in
terms of seasonality and how the annual cycle was dealt with, in
view of the new idea of growing maize. Of course once this idea
and practice was adopted by these prehistoric peoples, their entire
settlement-subsistence system began to change as they modified their
lifeways to cope with the ever greater importance that maize husbandry was acquiring, as these prehistoric peoples became evermore re-
liant upon it.

Seasonality and the Annual Cycle

The spatial distribution of components representing these early
agriculturalists in southwestern Ontario, southeastern Michigan and
northwestern Ohio (c.f. Map No.1) is such that most of them are located, in concentration, in environmental-ecologically similar regions.
With rare exception these components are located in environments of
riverine and/or lacustrine orientation in bottomlands, swamp-marshlands, floodplains and mudflats, which are climatically similar.

The few components that have been located away from aquatic
environments which are inland and interior oriented, are very small,
and have produced a paucity of cultural material suggesting that
they represent late fall through early spring encampments of these
early agricultural peoples who dispersed into the interior regions
during the cold and more vigorous times of the year to carry out
winter hunting activities.

Winters states (1969: 111) that "by settlement system we refer
to the functional relationships among a contemporaneous group of
sites within a single culture". Thus, to understand the seasonal
cycle of these early agriculturalists, the following quote is
offered. This quote pertains to the Ojibway and Cree, contemporary

hunting and gathering societies from northern Ontario. The socio-
economic level of integration of these people is probably somewhat
similar to that of these early agriculturalists.

> "Only in summer did the members of the
> band usually come together for any length
> of time . . . With the approach of fall,
> the people separated, each hunting group
> moving to its accustomed hunting area. . .
> Each hunting group appears to have been
> quite small, being composed of two to
> four closely related nuclear families
> numbering in all, ten to fifteen people,
> under the direction of the eldest male.
> The hunting group may have closely cor-
> responded to, or been identical with,
> the extended family" (Rogers 1963: 71).

The small scattered encampments, which have been located in the
interior regions are small, few in number (probably due to difficulty in locating them), and usually disclose very little cultural material. It is probable that these early agriculturalists moved away
from the reduced food availability and harsh effects of aquatic environments during the fall and winter to hunt in regions of the interior. As Conway notes (1975: 6):

> "If the winter section of the seasonal
> cycle included family units separated
> into hunting territories, the winter
> camps would be scattered. Also, re-
> covered winter tool kits can differ
> notably from the spring/summer arti-
> facts, perhaps to the point of being
> almost aceramic."

In this way these people could exploit the seasonally variable
resources such as nuts, berries, and game animals, which would not
be found in close proximity to water during the harsher period of
the year.

As such, the early agricultural peoples in southwestern Ontario,
southeastern Michigan and northwestern Ohio probably coalesced into
larger focal settlements which were aquatic riverine and/or lacustrine oriented in bottomlands, swamp-marshlands, floodplains and mudflats during the late spring, summer and early fall. At these times
of the year the spring fish runs could be exploited, as could wild
fowl and marsh plants. In a similar manner, small plots of maize
could be planted and looked after. Sites pertaining to these early
agricultural peoples, which have been located in these aquatic

environments are usually spatially less dispersed, they are larger,
and they are much richer in the amount of cultural material which
they disclose. The physical inability to be present in many of these
locales during the early spring due to flooding and ice rafting, not
to mention the damp and harsh nature of aquatic environments such
as these during the late fall through early spring, indicate that
these sites represent late spring through early fall encampments.
When all of these data are taken into consideration along with the
faunal and floral remains (which further indicate seasonality), the
late spring through early fall aspect of the seasonal cycle is further confirmed.

Such a view of the settlement-subsistence system of these early
agriculturalists conforms directly to what we already know about the
settlement-subsistence system of earlier Middle Woodland peoples in
the lower Great Lakes (Ritchie 1969; Fitting 1970; Stothers 1975d).

It is proposed that these rich, ecologically diverse areas in
southwestern Ontario, southeastern Michigan and northwestern Ohio
provided an ecological environment conducive to agricultural pursuit,
and that initially, limited agricultural pursuit was conducted on
an elementary basis in conjunction with the earlier, underlying Middle Woodland economic pattern of hunting and fishing. This idea is
comparable to Struever's theory of agricultural development in the
Illinois Valley (Struever 1968), except that a 'sui generis' theory
of agricultural development is not being postulated.

The Social Ramifications of a Settlement-Subsistence Shift

While the initial adoption of limited maize husbandry in various
regions of the Lake Erie drainage basin did not immediately alter the
earlier Middle Woodland economic, settlement or subsistence patterns,
it did establish a base from which the pre-agricultural levels of
socio-cultural integration (band societies of wandering hunters,
gatherers and fishers) would very soon be transformed to those levels
of socio-cultural integration attendant to sedentary village life.
Thus, in long range perspective, the adoption of maize husbandry,
by non-agricultural peoples had an impact that was to be 'revolu-
tionary'.

In pre-agricultural times in southwestern Ontario, southeastern
Michigan and northwestern Ohio, the subsistence basis was such that
people were not sedentary, but moved about periodically to exploit
seasonally variable natural resources. It was learned that an easily
controllable food resource was available in the form of maize, and
it was learned that it was essential to become at least seasonally
sedentary in order to be at hand for care and protection of the maize
crops. Soon people realized that surplus maize produce could be
dried and stored for the most devastating periods of the annual cycle.

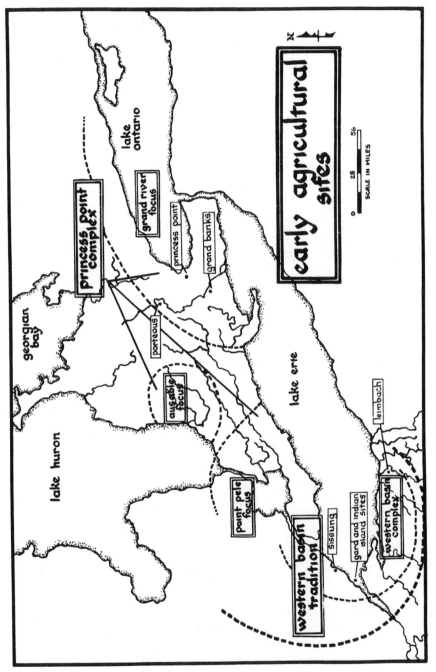

Map No. 1

Research conducted in southwestern Ontario (Stothers 1972, 1973 a, 1975b, 1976) indicates that the Princess Point culture, after they had learned about and began to practice maize agriculture, slowly developed larger and larger social aggregates. After A.D. 900 these aggregates shifted out of aquatic environments to establish permanent pallisaded villages on the summits of well drained, sandy hills (Stothers 1976).

CONCLUSIONS

It has been described in detail (Stothers 1976: 135-139, 154-167) how the advent of maize agriculture probably brought about the concept of matrilocality, permanent territoriality, endemic warfare, and a population 'explosion'.

The Princess Point Complex of southwestern Ontario represents a pre-Iroquoian manifestation which eventually gives rise to the neutral Iroquois Confederacy, while the Western Basin Complex and Riviere aux Vase peoples of southeastern Michigan and northwestern Ohio eventually gave rise to the later aspects of the Western Basin Tradition (Younge and Springwells phases), which was probably also an Iroquoian cultural group (Stothers 1975c). It would appear that an agricultural 'revolution', as in other areas of the world, was instrumental in attainment of ever greater levels of socio-cultural sophistication. As such, the introduction of maize agriculture into the region of the lower Great Lakes, along with the resultant settlement-subsistence shift from seasonally wandering bands of hunters, gatherers and fishers, to larger social aggregates of sedentary village agriculturalists, is viewed as establishing the social foundations for later Iroquois culture.

ADDENDA

1.) Since this paper was originally written, two new radiocarbon dates have been received for the Gard Island No.2 site. Another sample of human bone from Burial Feature 3a was processed and a resultant date of A.D. 850 ± 75 (DIC-418) was received. A sample of human bone from Burial Feature 6 was also processed and a date of A.D. 830 ± 125 (DIC-419) was received. When all of the radiocarbon dates for the Gard Island No.2 site are averaged (not including DIC-417, which is rejected as erroneous), a mid-eighth century temporal placement can comfortably be assigned to this component (Lozanoff and Stothers 1975: 3).

2.) It has recently been learned (Prufer and Shane 1972: 16; Shane: personal communication) that carbonized maize was recovered from a Late Woodland refuse pit (Feature 19) at the Leimbach site, which is located near the Vermillion River in Lorain County, Ohio (c.f. Map No.1). "This feature yielded two Vase Tool-Impressed rim sherds (from two vessels), one triangular projectile point, and wood char-coal which provided a date of A.D. 575 ± 180 years (GX-1743). Plant remains recovered by flotation have been identified by Richard Yar-nell, and include hickory nut shell, three chenopodium seeds, and corn kernels (Prufer and Shane 1972: 16)." Shane states in a personal communication (letter dated February, 1976) that ". . . corn agri-culture appears to have been practiced throughout the Late Woodland sequence in north-central Ohio." Based on the associated cultural material and radiocarbon date from Feature 19 at the Leimbach site, this maize is also associated with the early Late Woodland, Riviere aux Vase phase of the Western Basin Tradition.

ACKNOWLEDGEMENTS

The authors would like to extend their thanks and appreciation to the following people and institutions, without whose aid this article and its information content could never have been realized.

The University of Toledo is thanked for their continuing finan-cial aid and support, not only for processing radiocarbon samples, but also for aiding in financing the field excavation projects which have produced the raw data and material for carbon dates, faunal, floral, human skeletal, ceramic and lithic analyses.

Mr. Frank Huntley (Industrial Minerals Geologist at Libbey-Owens Ford Co.), and Mr. Larry Graves (Dept. of Geology, Bowling Green State University), are thanked for their aid in lithological and pedological studies and analyses, both in the field and in the lab-oratory. Dr. Irene Stehli (Director of the Radioisotopes Laboratory, Dept. of Geology, Case Western Reserve University), is thanked for her assistance in processing and interpreting radiocarbon assay re-sults. Dr. James E. Fitting (Chief, Cultural Resources Section, Commonwealth Associates Inc.) is thanked for his discussions con-cerning the Late Woodland Younge sequence and people of southeastern Michigan and northwestern Ohio. Dr. Fitting and the Dept. of Natural Resources of Michigan are acknowledged for their aid in carrying out research work on Michigan State land. Mr. Ken Reau (Manager, Erie Game and Shooting Preserve), extended his aid and hospitality to the senior author and his research team, in allowing prehistoric material to be studied and photographed, and in allowing excavation to be un-dertaken on the property of the Erie Game and Shooting Preserve. Mr. Walter Fedosik has been of monumental assistance in transporting personnel of various research teams to and from island and off-shore locations in Lake Erie.

Academic and theoretical content are in part the result of

extensive discussion and ceramic analyses with David Miller, Assistant Curator of Anthropology, Dayton Museum of Natural History.

Dr. James F. Metress (Dept. of Anthropology, University of Toledo), is thanked for reading and critically commenting on earlier drafts of this paper.

Of course the excavation and laboratory processing of prehistoric material could never have been accomplished without the aid of countless individuals, who are students or members of the ever reliable and helpful Toledo Area Aboriginal Research Club, (Inc.). Miss Jackie Royer (Dept. of Anthropology, University of Toledo), drafted the excellent map which accompanies this article. Catherine Stothers and Terese Flynn aided in proofreading and typing the final manuscript.

To all of these people we extend our sincere thanks. However, any error in, or shortcoming of this article is entirely the responsibility of the authors.

REFERENCES

Baerreis, David and Reid A. Bryson. 1965. Climatic episodes and the dating of Mississippian cultures. Wis. Arch. 46(4). (n.s.) Lake Mills.

Bryson, Reid and W.M. Wendland. 1967. Tentative climatic patterns for some late glacial and post-glacial episodes in central North America. Pages 271-298 in W.J. Mayer-Oakes, ed. Life, land and water. University of Manitoba Press. Winnipeg.

Brose, David S. and P. Essenpreis. 1973. A report on a preliminary archaeological survey of Monroe County, Michigan. Michigan Archaeologist 19(1-2).

Conway, Thor A. 1975. Salvage excavations of an Early Woodland component at the Schoonertown site. Canadian Archaeological Association - Collected Papers March 1975. Ont. Min. of Nat. Res., Hist. Sites Branch. Toronto. Research Report 6:1-7.

Cutler, Hugh and Leonard Blake. 1971. Maize from prehistoric Ontario Indian sites. Unpublished manuscript in D. Stothers' possession. 4pp.

Cutler, Hugh and Leonard Blake. 1973. Plants from archaeological sites east of the Rockies. Missouri Botanical Garden. St.Louis, Missouri.

Fitting, James E. 1965. Late Woodland culture in southeastern Michigan. Anthro. Papers of the Museum of Anthro. 24. Univ. of Mich. Ann Arbor.

Fitting, James E. 1966. Radiocarbon dating the Younge Tradition. Amer. Antiq. 31(5):738.

Fitting, James E. 1970. The archaeology of Michigan. Natural History Press. New York.

Griffin, James B. 1960. Climatic change: a contributing cause of the growth and decline of northern Hopewellian culture. Wis. Arch. 41(1):21-33.

Johnston, Richard B. 1968a. The archaeology of the Serpent Mounds site. Occasional Papers in Art and Arch. 10. Univ. of Toronto Press.

Johnston, Richard B. 1968b. Archaeology of Rice Lake, Ontario. Anthro. Paper 19. National Museum of Canada. Ottawa.

Jones, Robert W. 1931. The Clemson mound. Pages 89-111 in Pennsylvania Historical Commission, 5th Annual Report. Harrisburg.

Lozanoff, Scott and David M. Stothers. 1975. The Gard Island no.2 site: a biocultural analysis of an early Late Woodland population in southeastern Michigan. Toledo Area Aboriginal Research Bull. 4(3).

McCann, Catherine. 1971. Notes on the pottery of the Clemson and Book mounds. In B. Kent, Ira Smith III and C. McCann, eds. Foundations of Pennsylvania prehistory. Anthro. Series of the Pennsylvania Historical and Museum Commission. 1.

Prahl, Earl J. 1969. Preliminary comparison of three prehistoric sites in the vicinity of the western Lake Erie shore. Toledo Area Aboriginal Research Bull. 1(1):32-63.

Prufer, Olaf H. 1964a. The Hopewell complex of Ohio. Pages 37-83 in J. R. Caldwell and R. L. Hall, eds. Hopewellian studies. Ill. State Museum Scientific Papers 12(1).

Prufer, Olaf H. 1964b. The Hopewellian interaction sphere: theoretical implications of a prehistoric socio-religious system. Ohio Acad. of Sciences. Sec.1. Cleveland.

Prufer, Olaf H. 1964c. The Hopewell cult. Sci. Amer. 211(6):90-102.

Prufer, Olaf H. and Orrin C. Shane, III. 1972. The Portage-
Sandusky-Vermillion river region in Ohio. In David S. Brose, ed.
The late prehistory of the Lake Erie drainage basin: a symposium.
Scientific Papers of the Cleveland Museum of Nat. Hist. In press.

Ritchie, William A. 1938. Certain recently explored New York
mounds and their probable relation to the Hopewell culture. Research Records of the Rochester Museum of Arts and Sciences. 4.

Ritchie, William A. 1969. The archaeology of New York state.
Revised ed. Natural History Press. New York.

Rogers, Edward S. 1963. Changing settlement patterns of the Cree-
Ojibwa of northern Ontario. SW Journal of Anthro. 19(1):64-88.

Spence, Michael W. 1967. A Middle Woodland burial complex in the
St. Lawrence valley. Anthro. Paper 14. National Museum of Canada. Ottawa.

Stothers, David M. 1972. The Princess Point complex: a regional
representative of an early Late Woodland horizon in the Great
Lakes area. In David S. Brose, ed. The late prehistory of the
Lake Erie drainage basin: a symposium. Scientific Papers of the
Cleveland Museum of Nat. Hist. In press.

Stothers, David M. 1973a. Early evidence of agriculture in the
Great Lakes. Canadian Archaeological Association Bull. 5:62-76.
Ottawa.

Stothers, David M. 1973b. Radiocarbon dating the culture chronology of the western Lake Erie basin: part 1. Toledo Area Aboriginal Research Bull. 2(3):26-42.

Stothers, David M. 1975a. The Western Basin tradition: a preliminary note. Toledo Area Aboriginal Research Bull. 4(1):44-48.

Stothers, David M. 1975b. The archaeological culture history of
southwestern Ontario. Canadian Archaeological Association - Collected Papers March 1975. Ont. Min. of Nat. Res., Hist. Sites
Branch. Toronto. Research Report 6:116-127.

Stothers, David M. 1975c. The emergence and development of the
Younge and Ontario Iroquois traditions. Ont. Arch. 25:21-30.

Stothers, David M. 1975d. Middle Woodland manifestations in southwestern Ontario. Paper presented at Ontario Archaeological Society symposium 'Ontario Iroquois Prehistory'. Toronto, Ont. 19 pp.

Stothers, David M.   1975e.   Radiocarbon dating the culture chronology of the western Lake Erie basin: part 2.   Toledo Area Aboriginal Research Bull. 4(2):32-50.

Stothers, David M.   1976.   The Princess Point complex.   Mercury
   Series.   Archaeological Survey of Canada.   Ottawa.   In press.

Stothers, David M., David Miller, and Thomas Berres.   n.d.   The
   Gard Island no.3 and Indian Island no.3 sites: two early Late
   Woodland fishing stations in southeastern Michigan.   In prep.

Stothers, David M. and Earl Prahl.   1972.   The Squaw Island site
   (33-SA7), Sandusky County, Ohio.   Toledo Area Aboriginal Research
   Bull. 2(1):1-19.

Struever, Stuart.   1964.   The Hopewellian interaction sphere in
   riverine-western Great Lakes culture history.   Ill. State Museum
   Scientific Papers   12:85-106.

Struever, Stuart.   1968.   Woodland subsistence-settlement systems
   in the lower Illinois valley.   Pages 285-312 in L. and S. Binford,
   eds.   New perspectives in archaeology.   Aldine Press.   Chicago.

Struever, Stuart and K. Vickery.   1973.   The beginnings of cultiva-
   tion in the midwest-riverine area of the United States.   Amer.
   Anthro. 75(5):1197-1220.

Taggart, David W.   1967.   Seasonal patterns in settlement, subsistence, and industries in the Saginaw Late Archaic.   Michigan
   Archaeologist 13(4):153-170.

Vickery, Kent D.   1970.   Evidence supporting the theory of climatic
   change and the decline of Hopewell.   Wis. Arch. 52(2):57-76.

Winters, Howard D.   1969.   The Riverton culture: a second millenium
   occupation in the central Wabash valley.   Ill. State Museum Report 13.

Wright, James V.   1966.   The Ontario Iroquois tradition.   National
   Museum of Canada.   Bull. 210.   Ottawa.

Wright, James V.   1972.   Ontario prehistory: an eleven-thousand-
   year archaeological outline.   National Museum of Canada.   Ottawa.

Yarnell, Richard A.   1964.   Aboriginal relationships between culture and plant life in the upper Great Lakes region.   Anthro.
   Papers of the Museum of Anthro. 23.   Univ. of Mich.   Ann Arbor.

Yarnell, Richard A. 1973. Report on plant remains from the Porteous site (AgHb-1), Brantford, Ontario. Unpublished manuscript.
2pp. 1 chart.

Yarnell, Richard A. 1974. Report on archaeological plant remains
from ten sites in northwestern Ohio. Unpublished manuscript.
5pp. 1 chart.

Yarnell, Richard A. 1975a. Report on archaeological maize. Unpublished manuscript. 9pp.

Yarnell, Richard A. 1975b. Maize from archaeological sites on
Indian Island (1974). Unpublished manuscript. 3pp.

THE PLACE OF THE AMERINDIAN IN THE ORIGIN OF THE SOUTHERN APPALACHIAN GRASS BALDS

James F. Metress

The University of Toledo

Toledo, Ohio

ABSTRACT

An examination and critical reappraisal of the ecological and historical explanations concerning the origin of the grass bald areas of the Southern Appalachian summits. The possibility that the activity of Amerindians could have led to the formation of the balds as archaeological disclimax will be reexamined and proposals for future research will be suggested.

INTRODUCTION

In the past a number of researchers have discussed the Amerindian as an ecological factor (Day, 1953, Heizer, 1955, and Stewart, 1956). This paper will consider the role of the Amerindian in the origin of the grass balds of the southern Appalachian mountains. This issue has been considered by others in the past, but the present study represents a reorganization and synthesis of past and present ideas on the subject.

GRASS BALDS AS AN ECOLOGICAL PROBLEM

Grass balds are one of four discontinuous community types that are part of the vegetational pattern of the southern Appalachian mountains. They occur with few exceptions in the southern section of the Blue Ridge province (Fenneman, 1938) in western North Carolina and along the Tennessee-North Carolina line. The other discontinuous community types are the spruce-fir forest,

beech gap forest, and the heath balds (Whittaker, 1956). Grass
balds should not be confused with the heath balds which are different in species composition and micro-climate. The heath balds
appear to be of natural origin.

The Grass Balds are located at an altitude of approximately
3000 to 6300 feet on the most exposed summits extending a short
distance down the south and west slopes. They appear as grassy
meadows surrounded by forest. The meadows are dominated by herbaceous perennials, but shrubs are found in the forest edge community surrounding the meadow. The change from bald to forest is
abrupt without gradual thinning of trees. The bald areas are
restricted in size, varying from one-quarter acre to 100 acres,
and exhibit great diversity in vegetative composition and site
location.

The meadows are dominated by mountain oat grass (Danthonia
compressa) but sedges (Carex flexuosa, etc.) are common in the
moister areas of the bald. Herbaceous plants, such as Fragaria
Virginia (Strawberry), Potentilla canadensis v. Caroliniana (cinque-
foil), and Rumen acetosella (living sheep sorrel), are important
components of the meadow area.

The forest edge community is dominated by Ericads, such as
Vaccinium constablesei, pallidum and hirsutum, Rhodendron caten-
dulaceum, and Ligonia liquistrina. The community also contains
small trees, such as Betula lutea, Amelanchier laevis, Crataegus
macrosperma, Salix sp., and Rubus canadensis, and dwarfed trees of
the same type as the surrounding beech, oak, spruce and fir forests.

HISTORICAL ATTEMPTS AT AN EXPLANATION OF THE GRASS BALDS

Mitchell was the first to mention the grass balds in 1835,
when he described Roan Mountain Bald in North Carolina. Asa Gray
in 1841 made an extensive and valuable plant list for Roan Mountain Bald. However, he made no attempt to explain the Bald's
origin. The Michaux's, an early father and son team, and Bartram
both travelled through the Southern Appalachians during the 1700's.
These earlier explorers made no mention of the grass balds; but it
is well known that they followed the lowland trails.

In 1894, Edson made the first attempt to scientifically explain the origin of the balds. After spending the winter on Roan
Mountain she proposed that ice formation at certain places was so
great that it destroyed the woody plants and that grass came to
replace them. Harshberger in 1903 accepted Edson's theory and
added an exposure factor. In his study of the flora of the North
Carolina mountains he distinguished four ecological formations

largely determined by climate. These were the mixed deciduous
formations, coniferous forest, subalpine dwarf tree-shrub formations and the subalpine treeless formations. He felt that the
latter two were the result of winter ice storms.

Davis (1930) mentioned several possible causes of the balds,
such as grazing, fire and clearing and a failure of the orchard
association to replace its trees. He also mentioned that he knew
of "No legend or evidence of Indian occupancy," but a lack of legends may indicate a lack of interest on the part of the Indians.
The evidence for Indian occupancy has never been fully investigated. Furthermore, a number of Cherokee legends concerning the
mythological origins of the bald areas have been described by
Mooney (1898) and Wilburn (1950, 1952).

W. H. Camp (1931, 1936) sought to explain the balds by a local
climatic theory that emphasized exposure to the hot drying south-
westerly winds during the late summer in the Little Tennessee River
Valley. Although Camp observed only one bald area (Gregory Bald),
he suggested that balds were originally meadows, with ericaceous
islands with the ericads and grasses sharing dominance, maintained
by fire and grazing. He seems to have come to this conclusion
after talking to an elderly local informant who told him that the
old folks, "long dead," said they were blueberry meadows and always
balds.

Stanley Cain (1931, 1936) said that balds were of natural
origin. He felt that the soil was too rich and deep to have been
formed in recent history and decided that they must antedate the
appearance of the white man in the region. Camp (1931), from his
studies on Gregory Bald, disagreed with Cain with regard to the reported homogeneous richness and depth of the grass bald soil.
Mark (1958) found little difference in the grass bald soil and the
adjacent forest except that the bald soil was a little more acid
and the forest soil contained a little more organic matter. He
found that the soil differences between balds were actually greater
than between the balds and their adjacent forests.

In 1931, P. M. Fink who did not propose any theory of origin
did state that he felt an Amerindian explanation was unlikely. He
thought the task was too great given the Indian's limited technological level, but Day (1953), and others have documented the
Indian's ability to clear large areas of forest in other parts of
the east. Fink (1931) also doubts that, even if they had had the
ability to clear the areas, they would have expended such energy
without some vital reason. All of these attempts to predict the
value orientations of a prehistoric people, based on present standards or conditions, borders on pseudopsychological archaeology.

Frederick Clements (1936) said the balds were produced and maintained by periodic burning, but he found few if any backers for this view. If fire alone could produce the mountain grass balds, then vast areas of the Southern Appalachians would have been covered by grass balds. Gates (1941) suggested that gallflys could have killed off the dominant oaks and allowed mountain grasses to assume dominance. A theory of disease selection such as this based on an insect vector hardly seems plausible and as far as I know is without precedent. Peattie (1936) explained the balds on the basis of excessive evaporation due to the isolation of the peaks, but the balds are no more isolated than the forested peaks.

In 1941, Brown rejected the Amerindian theory because he believed that the alleged lowland habits of the Indian negated it. He also felt tree succession was proceeding very slowly or was temporarily halted due to climatic change.

Bertram Wells, a long time student of the grass balds, is the leading proponent of an Amerindian origin (1936, 1938, 1946, 1956, 1963). Wells has visited more grass balds than any other researcher and has been the most persistent spokesman for the Amerindian theory. On the other hand, Whittaker (1956), in his landmark study of the vegetation of the Smoky Mountains has attempted a biological explanation. But Wells (1956) in turn has rejected Whittaker's proposals on several solid points which will be discussed later.

Mark and Billings (Billings 1951, Billings and Mark 1957, Mark 1958) proposed that the grass balds are related to the formation of a grass bald susceptible zone 4,000 years ago. They feel that during the post-Wisconsin Xerothermic fluctuations the relative positions of the red oak forest, transition zone, and balsam spruce forest moved upward about 300 to 1,000 feet. The authors hypothesize that a susceptible zone was created around 4,000 years ago when a slow cooling of climate through a very narrow zone less than a half mile wide killed off the red-oak forest, thus allowing the wild mountain oat grass to take over the area.

It would seem if Mark and Billings ideas were correct there should be a much greater area of the Southern Appalachians covered by grass bald vegetation. But the recorded grass balds are relatively few in number and rather small in area. Also, if the grass balds are relicts of once more extensive grassy areas, how have they been able to hold out against the usual forest succession for 4,000 years, while most of the area is covered by unbroken forest?

NATURAL ORIGINS THEORIES AND THEIR SHORTCOMINGS

Of the natural origin theories those of Edson (1894), Camp (1931, 1936), Cain (1931, 1936), Clements (1936) and Gates (1941) are rejected for the reasons mentioned in my historical survey of the literature in this paper. On the other hand the ideas of Whittaker (1956), Mark (1958), Billings (1951), Billings and Mark 1957) and Brown (1941) will be considered further at this point.

Whittaker in his monograph Vegetation of the Great Smoky Mountains in 1956 proposes the following explanation. He believes that climatic factors such as low temperature and low moisture associated with south and west exposure are critical. He further postulates that soil water loss from runoff on the slopes and the desiccating effects of the full force of the winds blowing from the west without any obstructions creates a moisture gradient that excludes trees. The trees involved are high altitude variants, of dwarfed stature and located probably beyond the most favorable growing conditions, such as temperature, length of growing seasons and soil fertility.

The Whittaker explanation falls short in a number of areas. First, if this "extreme" exposure idea were correct one would expect to find most of the Appalachian summits in grass balds. But in fact the pre-White grass balds are few in number, small in size and scattered irregularly along the ridge tops. Secondly, Whittaker's idea cannot account for the abrupt boundaries between the forest and the grass bald, since extreme exposure would not be so abrupt. Thirdly, the bald vegetational association is also characteristic of the high ridge Indian trails which due to forest protection on each side negate an extreme exposure hypothesis.

Mark (1958) has proposed a climatic theory for the origin of the balds. He believes that the post-Wisconsin Xerothermic fluctuation caused the ecotone (Zone of transition), between the Northern Red Oak and balsam-spruce forest, to shift upward 300 to 1,000 feet higher than its present position. When subsequent cooling occurred, a "bald susceptible" zone was formed along the ecotone. This could be due to the destruction of either the deciduous forest near its upper limit or the spruce-fir forest near its lower limit. The removal of the protective forest canopy could have led to a changed microenvironment in which the tolerance limits of the tree seedlings would be exceeded. This would make reinvasion from above or below either very slow or impossible.

Open area bald environments would be more severe for seedlings with respect to wind, drought, temperature extremes, snow drifting, etc. Destruction or reduction of forest tree populations would lead to lack of seed sources that would also impede reinvasion. It is also possible that deer and bison may have helped to maintain them before the introduction of domestic stock and European farming practices.

The extermination of trees by temperature over such a narrow zone ($\frac{1}{2}$ mile) seems unlikely. As yet no one in plant ecology has been able to demonstrate that climatic cooling can stop the normal woody succession. Also, even if one concedes such an unlikely change could occur, there should be thousands of grass balds located in the Appalachian transition zone, which of course is not the situation.

Since the Appalachian grass balds occur in the ecotone zone, Mark and Billings (Mark, 1958, Billings and Mark, 1957) feel they are a successional type that will disappear. But since Mark's (1958) cooling period occurred approximately 4,000 years ago one would think such small relict areas would have succumbed to the natural processes.

Gersmehl (1969) rejects the ecotonal hypothesis on two grounds. First, states Gersmehl, there is ample historical evidence that balds were located above the spruce forest limits. Secondly, present observations indicate that the absolute tolerance limits of the spruces and oaks overlap to a considerable extent. He also demonstrates that due to the wide altitude range for the ecotonal zone, either forest type would be capable of much expansion beyond the present transition zone between them. Gersmehl (1971) proposes a theory of limited seed dispersal range as an explanation for the delayed reforestation of the grass balds, and presents experimental evidence that negates soil moisture, insolation and wind velocities as significant factors in bald maintenance. While these studies seem to render the "ecotonal hypothesis" and other ecological theories inadequate, they still do not explain the "origin" of the balds, only their maintenance.

Brown (1941) gave a rather complete idea of the structure and composition of a grass bald community on Roan Mountain. He felt that insufficient moisture prevented seedlings from competing with grasses in the past but due to recent climatic changes which shortened the dry periods the balds are being invaded by shrubs and conifers. In 1953, Brown suggested that a lack of seed sources or factors that inhibit early seed development were responsible for the relative slow invasion of the balds. Gersmehl's studies (1969, 1970, 1971) would seem to support the idea of lack of suitable seed sources and poor seed dispersal. Radford (1968) agrees with Billings and Mark (1957) that the balds occupy potential spruce-fir zones and that competition of the grasses with tree seedlings under the environmental conditions of the bald is detrimental to the seedlings. But still he does not explain the origins, only maintenance or bald persistence.

THE AMERINDIAN ORIGIN THEORY

Wells (1936, 1937, 1938, 1946, 1956, 1963) has proposed that the grass balds are the result of Indian occupation and therefore considers them archaeological disclimaxes. Iverson (1941) and Godwin (1944) have suggested archaeological disclimax as an explanation for heath land communities in Denmark and East Anglia, England.

Wells further (1937) feels that the balds could possibly represent three types of occupational sites: campsites, game lures, and lookouts. Campsites could have been located on gentle slopes, on round ridge tops or gaps near high springs. Wells postulates that the steeper slopes could have been the site of game lures. They were smaller, but deer and turkey could have been shot on them very effectively from the surrounding forest. Lookouts were small, located on or near sharp summits and were characterized by excellent views or large areas of wilderness.

The campsites were probably summer camps. They were needed because ascent and descent in one day would leave little time for hunting. According to Truett (1935:6), the Indians preferred ridge trails to stream side trails for hunting since there was less undergrowth, fewer streams to cross and they offered a better opportunity to sight game and enemies.

Springs are infrequent in the high summits of Southern Appalachia, so it is probably not coincidence that over half the balds identified are associated with a spring. It is possible that some balds were meeting places and in the case of some of the larger balds refuge areas in times of hostilities.

The major objections to the Amerindian theory are largely couched in terms of technological capability and motivation for such an expenditure of energy. But decades of use and slow destruction of trees and shrubs for fire wood could lead to ever enlarging open areas. The stone axe of the Indian was a much more efficient tool than ethnocentric authors of today are able to perceive. Many of the grass balds may have had their origins as an expanded trail or an area where a number of trails criss-crossed. It is also possible that they could be modified heath balds.

The evidence for an Amerindian origin for the grass balds is at this time largely inferential since no one has systematically surveyed or excavated a grass bald (Coe, 1974). This inferential evidence has both negative and positive aspects. Negatively, the absence of grass balds from most of the high mountain acreage would seem to contradict any "natural" explanations. This evidence may be the most important.

On the positive side there are a number of important observations that seem to support an Amerindian origin. Old trails (100
yards or so wide) in the same area are characterized by the same
vegetation complex as the grass balds. Either the grass balds are
expansions of the same processes that produced the trails or one
must postulate the same natural factors that produced the balds
operated lineally without effecting the forest on either side.

Some of the balds near the springs show evidence of disturbed
fire areas that have been lowered by the washing of rains. At
Andrews Bald, and Spence Field Bald, Swain County, North Carolina,
there are semi-circular areas that are eroded to a depth of a foot.
These areas are invariably found near the "bald" spring.

Dr. L. A. Whitford of North Carolina State University noted
three circular depressions in the granite on the lower side of
Andrews Bald. These depressions are suggestive of rock mortars
for grinding grain or seed (Wells, 1938). Also, a Smoky Mountain
National Park Naturalist, George Stupke, informed Wells (1937)
that an old mountaineer told him that his grandfather as a boy
heard the Indians claim that Mt. Sterling Bald, Haywood Center,
N. C., was made for the purpose of shooting game.

Balds or open areas of more recent origin exhibit the changes
leading to a mountain-oatgrass community. For example, a recluse
cleared a 5 acre mountain top for farming near Mountain Lake, Virginia. The area was abandoned in 1900 and was taken over by oat-
grass and held against the surrounding red oak forest until the
1950's when it was last observed. It would seem, if a cultural
process produces these results today in this region that it would
have been equally responsive to Amerindian activity.

If fire were necessary to maintain them, most balds would have
disappeared readily since most show little evidence of burning.
Also, in other areas of the Appalachians, thousands of acres of
burned areas have gone through the standard successional steps,
characterized by choke cherry thicket followed by regeneration of
the balsam-fir climax. But nowhere in the Southern Appalachians
is there evidence that fire has led to a typical grass bald community.

It would appear that interference was at the soil level.
Tramping of human feet and activity made normal succession impossible and when the area was finally abandoned it developed into a
mountain-oatgrass community. Mountain-oatgrass (Danthonia compressa), once established, can (through root-competition) prevent
any return of the woody plants. Liddle (1975) has recently reviewed the literature on the ecological effects of human trampling
on ecosystems and presents evidence from a number of studies that

indicate human trampling is capable of altering the normal successional processes. Bates (1935) carried out systematic and experimental studies of the effects of trampling. He observed that tread-
ing had direct mechanical effects on the vegetation and indirect
effects due to soil changes. There seems to be a reduction of the
total cover and the number of species in areas of heavy human activity and some plants appear to be more resistant to the mechanical effects of trampling. Numerous ecologists have noted that
soil compaction due to human trampling (Liddle 1975) affects the
germination and establishment of certain plant species.

These observations plus the physical fact of their location
on warm gentle slopes, near springs ideal for camps or hunting
areas, or on steep slopes with commanding views would seem to
press for at least the strong possibility of an Amerindian origin.

SUMMARY

It would seem from a review of the evidence that the various
ecological explanations for the origin of the grass balds are insufficient. Although some of the theoretical processes suggested
quite possibly contributed to the persistence of the balds, none
of these explanations shed light on the origin of the balds. It
is also apparent that the indirect inferential evidence for Amerindian activity as their source is strong and thus suggestive of a
human disclimax. But since little direct evidence of Amerindian
occupancy is available it is probably imperative that some of the
balds be systematically collected and excavated in an effort to
prove or disprove the initial hypothesis of B. W. Wells.

BIBLIOGRAPHY

Bates, C. H. 1935. The vegetation of footpaths, sidewalks, cart
    tracks and gateways. J. Ecol. 23:470-487.

Billings, W. D. 1951. Vegetational zonation in the Great Basin
    of Western North America, pp. 101-122. In: C. R. Colloque,
    Sur les bases ecologiques de la regeneration de la vegeta-
    tion des zones arides. UISB, Paris.

Billings, W. D. and A. F. Mark. 1957. Factors involved in the
    persistence of montane treeless balds. Ecology, 38:1:140-142.

Brown, D. M. 1941. Vegetation of Roan Mountain, a phytosoci-
    ological and successional study. Ecological Monog., 11:61-97.

Brown, D. M. 1953. Conifer transplants to a grassy bald on Roan
    Mountain. Ecology, 34:614-617.

Bruhn, M. E. 1964. Vegetational succession on three grassy balds
    of the Great Smoky Mountains. Ms. Thesis, Univ. of Tennessee,
    Knoxville.

Cain, S. A. 1931. Ecological Studies of the vegetation of the
    Great Smoky Mountains of North Carolina and Tennessee.
    Bot. Gaz., 91:22-41.

Cain, S. A. 1936. Ecological work on the Great Smoky Mountain
    region. J. South Appalachian Bot. Club, 1:25-32.

Camp, W. H. 1931. The grassbalds of the Great Smoky Mountains of
    Tennessee and North Carolina. Ohio J. of Sci., 31:157-164.

Camp, W. H. 1936. On Appalachian Trails. J. N. Y. Bot. Garden,
    37:249-265.

Clements, F. 1936. Nature and structure of the climax.
    J. Ecology, 24:252-284.

Coe, J. 1974. Personal communications.

Colton, H. E. 1859. The scenery of the mountains of western North
    Carolina and Northwestern South Carolina. W. L. Pomeroy,
    Raleigh, North Carolina.

Davis, J. H. 1929. Vegetation of the Black Mountains of North
    Carolina. Ph.D. Dissertation, Univ. of Chicago.

Davis, J. H. 1930. Vegetation of the Black Mountains of North Carolina: an ecological study. J. Elisha Mitchell Sci. Soc., 45:291-318.

Day, G. M. 1953. The Indian as an ecological factor in the Northeastern forest. Ecology, 34:329-346.

Edson, H. R. 1894. Frost forms on Roan Mountain. Pop. Sci. Mo., 45:30.

Fenneman, N. M. 1938. Physiography of the Eastern United States. McGraw-Hill, N. Y.

Fink, P. M. 1931. A forest enigma. Amer. Forests, 37:538-556.

Gates, W. H. 1941. Observations on the possible origin of the balds of the Southern Appalachians. Louisiana St. Univ. Press, Contributions to Zoology, 53:1-16. Baton Rouge.

Gersmehl, P. J. 1969. A geographic evaluation of the ecotonal hypothesis of bald location in the southern Appalachians. Proc. Assn. Amer. Geog., 1:51-54.

Gersmehl, P. J. 1970. A geographic approach to a vegetation problem: The case of the Southern Appalachians. Ph.D. Dissertation, Univ. of Georgia, Athens.

Gersmehl, P. J. 1971. Factors involved in the persistence of Southern Appalachian treeless balds: an experimental study. Proc. Assn. Amer. Geog., 3:56-61.

Gilbert, V. C., Jr. 1954. Vegetation of the grassy balds of the Great Smoky Mountains National Park. M.S. Thesis, Univ. of Tennessee, Knoxville. 73 pp.

Godwin, H. 1944. Neolithic forest clearance. Nature, 153:511.

Godwin, H.   1944. Age and origin of the "Breechland" heaths of East Anglia. Nature, 154:6.

Gray, A. 1842. Notes on a botannical excursion to the mountains of North Carolina. Amer. J. Sci. Arts, 42:44-47.

Guyot, A. 1863. Notes on the geography of the mountain district of western North Carolina. Hist. Rev. 15:251-318.

Harshberger, J. W. 1903. An ecological study of the flora of mountainous North Carolina. Bot. Gazette, 36:368-383.

Heizer, R. F. 1955. Primitive Man as an Ecological Factor.
    Kroeber Anthrop. Soc., 13:1-31.

Iverson, J. L. 1941. Denmarks Stenalder (land occupation in
    Denmark's Stone Age) Denmarks Geologiske. Undersogelse II,
    Raekke, 66:68.

Johnson, L. N. 1888. A tramp in the North Carolina Mountains.
    Bot. Gazette, 13:269, 318.

King, P. B. and A. Stupka. 1950. The Great Smoky Mountains.
    Sci. Monthly, 71:31-43.

Korstian, C. F. 1937. Perpetuation of spruce on cutover and
    burned lands in the higher Appalachians. Ecol. Monog.
    7:125-167.

Lanman, C. 1849. Letters from the Allegheny Mountains. Geo. P.
    Putman, New York.

Liddle, M. J. 1975. A selective review of the ecological effects
    of human trampling on natural ecosystems. Biol Conserv.,
    7:17-36.

Mark, A. F. 1958. An ecological study of the grass balds of the
    southern Appalachian Mountains. Ph.D. Dissertation, Duke
    University, Durham.

Mark, A. F.   1959. The flora of the grass balds and fields of
    southern Appalachians. Castanea, 24:1-21.

Mitchell, E. 1835. Notice of the height of mountains in North
    Carolina. Am. J. Sci. Arts, 35:378.

Mooney, J. 1898. The Cherokee Indians. 19th Rept. U.S. Bureau
    of Ethnology, Washington, D. C.

Olmsted, F. L. 1907. A journey in the back-country in the winter
    of 1853-54. V. II. G. P. Putman, New York.

Peattie, R. 1936. Mountain geography. Harvard Univ. Press,
    Cambridge.

Peattie, R.   1943. The Great Smokies and the Blue Ridge.
    Vanguard Press, New York.

Radford, S. W. 1968. Factors involved in the maintenance of the
    grassy balds of the Smoky Mountains National Park. M.S. Thesis,
    University of Tennessee.

Ramseur, G. S. 1960. The vascular flora of high mountain communities of the southern Appalachians. J. Elisha Mitchell Sci. Soc., 76:82-112.

Redfield, J. H. 1879. Notes of a botannical excursion into North Carolina. Torrey Bot. Club, 6:331-339.

Safford, J. M. and J. M. Killebrew. 1900. The elements of the geology of Tennessee. Foster and Webb Publ., Nashville, Tenn.

Schofield, W. B. 1960. The ecotone between spruce-fir and deciduous forest in the Great Smoky Mountains. Ph.D. Dissertation, Duke University, Durham, N. C.

Sharp, A. J. 1941. The Great Smoky Mountain National Park an important botannical area. Chronica Botannica, 7:296-297.

Stewart, O. C. 1956. Fire as the First Great Force Employed by Man. In: Man's Role in Changing the Face of the Earth. W. L. Thomas, Jr., ed. Univ. Chicago, Chicago, pp. 278-303.

Truett, R. B. 1935. Trade and travel around the Southern Appalachians before 1830. North Carolina Univ. Press, Chapel Hill.

Webb, L. J. 1964. An historical interpretation of the grass balds of Munya Mountains, South Queensland. Ecology, 45:1:159-162.

Wells, B. W. 1936. Origins of Southern Appalachian grass balds. Science, 83:283.

Wells, B. W. 1936. Andrews Bald: The problem of its origin. Castanea, 1:59-62.

Wells, B. W. 1937. Southern Appalachian grass balds. J. Elisha Mitchell Scientific Soc., 53:1:1-26.

Wells, B. W. 1938. Southern Appalachian grass balds as evidence of Indian occupation. The Arch. Soc. North Carolina, 5:2-7.

Wells, B. W. 1946. Archaeological disclimaxes. J. Elisha Mitchell Sci. Soc., 62:1:51-53.

Wells, B. W. 1956. Origin of Southern Appalachian grass balds. Ecology, 37:591.

Wells, B. W. 1963. The Southern Appalachian grass bald problem. Castanea, 26:98-100.

Whittaker, R. H.   1948.   A vegetational analysis of the Great Smoky
    Mountains.   Ph.D. Dissertation, Univ. of Illinois, Urbana,
    Illinois.

Whittaker, R. H. 1956. Vegetation of the Great Smoky Mountains.
    Ecological Monographs, 26:1-80.

Wilburn, H. C.   1950.   Cherokee landmarks around the Great Smokies.
    The Stephans Press, Asheville, N. C.

Wilburn, H. C. 1952.   The Judaculla place names and the Judaculla
    tales.   Southern Indian Studies, 4:22-26.

Yard, R. S.   1942.   The Great Smoky Wilderness.   Living Wilderness,
    7:14-17.

Zeigler, W. G. and Crosscup, B. S.   1883.   The Heart of the
    Alleghenies or Western North Carolina.   Alfred Williams.
    Raleigh, N. C.

SUSPENDED SEDIMENT AND PLANKTON RELATIONSHIPS IN MAUMEE RIVER AND MAUMEE BAY OF LAKE ERIE

Charles E. Herdendorf, David E. Rathke, Donna D. Larson
and Laura A. Fay

Center for Lake Erie Area Research
The Ohio State University, Columbus

ABSTRACT

In mid-March 1975, a survey was conducted in the lower Maumee
River, Maumee Bay and the adjacent portions of western Lake Erie to
determine the quality of water issuing from the river, particularly
suspended sediment, and the plankton populations in the receiving
body of water. The study was conducted during spring run-off in an
attempt to quantify the effect of the river discharge on such populations. The sediment-ladened effluent from the river coupled with
wave-resuspended sediments produced highly turbid water for approximately seven miles offshore at the outer terminus of the bay. Phytoplankton populations appeared to have been greatly inhibited in the
river and inner bay; they reached the greatest density at the outer
terminus of the bay and then decreased lakeward even though the clarity of the water improved. Cold Detroit River flow (1°C) influenced
the lakeward area and probably inhibited algal growth.

INTRODUCTION

Suspended sediments in the lower Maumee River and Maumee Bay
originate from at least three sources: (1) land run-off, (2) resuspension of bottom sediments and erosion of shore material by wave
action and (3) vessel operation, including dredging. Because the
Maumee River is one of the major contributors of sediment to Lake
Erie (Figure 1), the high level of suspended particles in the lower
river and bay has been suspected of having a profound effect on
plankton populations in these waters. Reid and Wood (1976) observed
that the lower courses of large streams flowing near base level
(such as the Maumee River) carry considerable loads of silt and

247

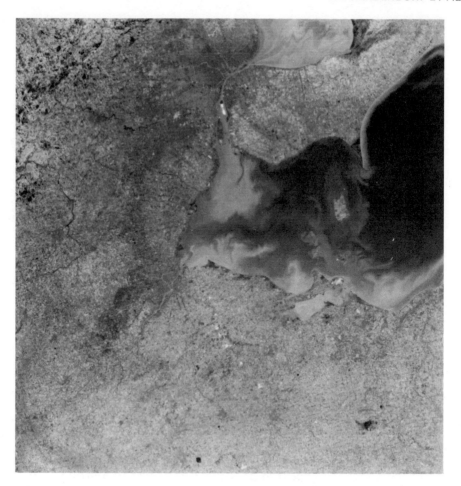

Figure 1.   ERTS -1 image (band 5) of western Lake Erie
             on 27 March 1973, showing sediment plumes issu-
             ing from the Detroit and Maumee Rivers during
             spring run-off.

clay particles which result in a decrease in phytoplankton due to
rapid quenching of light. In estuarine situations they found one
of the major effects of high turbidity was the reduction of light
penetration, thereby inhibiting photosynthesis and the production
of plants.

     To assess the magnitude of this effect measurements of suspended
material and plankton concentrations were made at several stations in
the river and bay and in the adjacent portion of Lake Erie during the
period of spring runoff in March 1975. Other factors influencing
the abundance of phytoplankton, such as temperature, water quality,

concentration of nitrogen, phosphorus and silica compounds, and zooplankton predators, were also investigated.

Maumee River Characteristics

The Maumee River is formed in Fort Wayne, Indiana by the merger of the St. Joseph River and the St. Marys River. The St. Joseph River originates in Hillsdale County, Michigan and flows southwest to Indiana. The St. Marys River originates in Shelby County, Ohio and flows northwest to Indiana. The Maumee flows from Fort Wayne through Defiance to Toledo and Lake Erie. The entire drainage basin is 6,586 square miles, 1,260 are in Indiana, 470 in Michigan and 4,856 in Ohio (Ohio Division of Water, 1960). The basin has a circular shape with a diameter of roughly 100 miles. The average gradient of the Maumee River is 1.3 ft/mile. The St. Marys averages 2.8 ft/mile and the St. Joseph 1.6 ft/mile. Some of the headwater tributaries have gradients as high as 10 ft/mile.

The main stem flows generally northeast from Ft. Wayne to Toledo, Ohio, about 135 miles distance. The Maumee River empties into Maumee Bay, a shallow basin at the southwestern tip of Lake Erie. The relatively flat basin yields a low gradient and a correspondingly sluggish flow. With a mean discharge of approximately 4,700 cfs (ranges from a high of 94,000 cfs to a low of 32 cfs), it is not a large river, but it is the largest tributary to the Great Lakes (Great Lakes Basin Commission, 1975). Mean flow for October 1974 through September 1975 was 5,420 cfs. The maximum flow occurred in February when discharge reached 49,000 cfs. The Maumee River accounts for only 3 percent of the flow into Lake Erie, but included in this discharge is 1.2 million tons of suspended solids annually, representing 37 percent of the total sediment load to the lake. Low relief, gentle gradient and fine-grained soils account for many of the rivers traits: its low velocity, muddiness and sediment-clogged bed (Horowitz, *et al.*, 1975). Floods occur annually during the early part of the year usually in February, March or April. The floods are caused by rainfall, frozen ground and melting snow. These factors are accentuated by the inability of the slow, sluggish Maumee to accept the increased load.

The lower 15 miles of the Maumee River can be considered a freshwater estuary. The formation of this estuary on Lake Erie is the result of a series of geologic events related to Pleistocene glaciation. The flow of the Maumee River was reversed from its southwest direction when the glacial lakes drained from the Erie Basin as the ice sheet melted exposing the Niagara River outlet. Base-level lowering accelerated river velocities and the Maumee valley was cut deeply into lacustrine deposits, glacial tills and bedrock. With the weight of the ice removed, the outlet eventually rebounded and produced a rise in water level. The lake encroached

upon the valley forming the present drowned stream mouth which is
analogous in many ways to a marine estuary. Virtually all of the
tributaries entering Lake Erie on the Ohio shore have estuarine-
type lower reaches where lake water masses affect water level and
quality for several miles upstream from traditional mouths (Brant
and Herdendorf, 1972).

    The estuary of the river begins just above the Maumee-Perrys-
burg Bridge where the bedrock riffles end. As the water enters the
estuary, its velocity abruptly dimenishes, except during major run-
off events, causing sedimentation of suspended particles. Within
the estuary, currents are extremely unstable. Reversals of flow
due to fluctuations in Lake Erie water levels have been measured by
Herdendorf (1970). The estuary is approximately a mile wide at
Eagle Point and nearly 30 feet deep in the dredged navigation channel. Early maps and charts show that the estuary was frequently
25 feet deep even before the Corps of Engineers began to improve the
harbor. Horowitz, *et al.*, (1975) considered the estuary a reservoir,
a sloshing dilution basin where the river is progressively mixed
with backflow from the lake, and a large settling basin where solids
from upriver are sedimented and occasionally scoured during periods
of major flushing. They also reported that the river water can be
relatively stagnant for long intervals.

    Miller (1968) observed that currents in Toledo Harbor exhibit
some of the properties of tidal currents, in that they reverse when
the water level changes from "flood" to "ebb" during wind tide and
seiche activity. During the period May to November 1966 he found
that 90 percent of the time the currents were less than 15 cm/sec
(0.5 ft/sec), and that the maximum speeds, about 45 cm/sec (1.5 ft/
sec), occurred during the greatest rate of change in water level,
whereas the minimums are at times of high and low water. A comparison of simultaneous data obtained from current meters and drogues
showed that current speeds in the mid-channel were up to 2.5 times
greater than near the channel edge. Miller also made estimates of
river discharge magnitude in relation to its effect on the currents
during seiching. Below 7,000 cfs current maintains its reversing
characteristics and the effect of river discharge on current speed
is not easily recognized. Above this discharge value the current
reverses from its down-channel direction only during periods of
rapid rises in lake level. For discharge rates greater than 20,000
cfs, the up-channel current component usually disappears even with
40 cm (1.3 ft) seiche amplitudes. Periods with this rate of flow
are infrequent and of short duration. Horowitz, *et al.* (1975) also
studied the hydraulic complexities of the estuary. In May and September 1974 they demonstrated periods of stagnation, river flushing
and reverse flow with stage recorders and drogues; as the water level
rose, lake water was pushed into the estuary, as it fell, river water
flowed into the bay. They estimated that each one-foot change in
water level causes the volume of water in the estuary to adjust by
approximately 120 million cubic feet.

Maumee Bay Characteristics

Maumee Bay lies at the western end of Lake Erie between 41°41'N and 41°45'N latitude and 83°20'W and 83°29'W longitude, mainly in Lucas County, Ohio. It is separated from Lake Erie by two spits: (1) Woodtick Peninsula, with North Cape at its southern tip, extends southerly from the Michigan shoreline and (2) Cedar Point projects northwesterly from the Ohio shore. Bathymetrically, Maumee Bay is a broad, shallow shelf sloping gently downward toward the northeast. The maximum depth is 10 feet below low water datum (LWD) and the mean depth is 5 feet (Benson, 1975). Relief of the bay floor is low except for the areas surrounding the navigation channel, which bisects the bay in a northeast-southwest direction. Adjacent to the channel, about 2000 feet from either side, are a series of linearly arranged islands and shoals, sandy at their surface (Charlesworth, 1974), that were formed from spoil banks when the channel was dredged to 25 feet in the 1930's. In the 1960's the channel was deepened to 28 feet. The navigation channel is now 500 feet wide and maintained to minimum depth of 28 feet below LWD. Dredging activities for the channel extend upstream 6 miles from the mouth of the river and lakeward for a distance of 17 miles. Presently, the dredged material from the inner five miles in the bay and all of the river is dumped in a diked disposal facility (Toledo Island) within the bay (a new disposal area on the east side of the river mouth is presently under construction) which will meet the dredged disposal needs for the next ten years). Lakeward from the five mile limit, the dredgings are dumped in designated areas in the open lake (U.S. Army, Corps of Engineers, 1974). At the entrance to the bay, Turtle Island on the north side of the channel and Cedar Point spit to the southeast also produce noticeable changes in the bottom topography. Maumee Bay and the adjacent portion of Lake Erie under consideration in this study covers approximately 30 square miles, seven square miles of which have sand deposits with the remainder composed of silt and clay.

Maumee Bay is characterized by a low clay shore, highly developed as a residential area on the west, and grading through a less intense development on the south to marsh on the northeast. Except for short reaches of sand on the bay side of Cedar Point, the Bay has practically no beaches. The material offshore is lacustrine clay with a thin overburden of recently deposited silt, except near Cedar Point where the overburden is a relatively thick layer of sand. The lacustine clay, up to 30 feet thick, was laid down in the glacial lakes which once covered a large part of northwest Ohio and southeastern Michigan. The lake clay is in turn underlain by sandy glacial till approximately 80 feet thick with Silurian dolomite below (U.S. Army Corps of Engineers, 1945). Slopes in the Maumee Bay nearshore zone are gentle. Within 1000 feet of the bay shore depths are generally less than 5 feet below LWD. Benson (1975) found that within 100 feet of the shore slopes ranged from 185 to 370 ft/mile, but lakeward of 500 feet offshore, slopes were generally less than 10 ft/mile.

The only sand deposits of any importance in Maumee Bay lie
between Cedar Point and Turtle Island in a modified spit extending
northward from Cedar Point and deposited by littoral currents from
the southeast. The Ohio Department of Natural Resources, Division
of Shore Erosion (Verber, 1954 and Hartley, 1960) extensively studied this deposit and concluded that none of the sand can come
from the Maumee River because the river loses its sand carrying capacity upon reaching lake level, near Perrysburg, as evidenced by the
fact that "muck" is found as the bottom material in the lower reach
of the river, at its mouth, and all the way out to the sand spit in
the bay. They also concluded that sand does not come from the shores
of Maumee Bay because the banks are composed of clay containing a
negligible amount of sand.

Walters and Herdendorf (1975) demonstrated that the Maumee
River has produced a sediment plume that extends from Maumee Bay
into western Lake Erie by measuring the concentration and distribution of mercury in the surficial sediments. This technique indicates
that recent sedimentation in Maumee Bay ranges from 0.5 to 1.0 cm/
year.

McBride (1975) lists seven sources of sediment supply to and
within Maumee Bay: (1) Maumee and Ottawa Rivers which supply fine
grained material to the western and southwestern part of the bay,
(2) Lake Erie which may supply additional fine grained material to
the bay, (3) relatively coarse material transported into the bay from
the north by the southeasterly flowing longshore current along the
lakeward side of North Cape, (4) coarse grained material brought into
the bay system by the longshore current flowing in a northwesterly
direction along Cedar Point, (5) residual coarse material found in
the southern bay as a result of the winnowing and removal of the
fine-grained component of the underlying Pleistocene, pebbly till-
clay, (6) lateral dispersal of coarse material from the spoil banks
along the navigation channel probably supplies some coarse material
to the southern bay, and (7) erosion of the rip rap along the southern shore and Point Place may supply a minor amount of sediment to
nearshore areas. He also divides the bay into three basic areas
based on the prevalent energy conditions and subsequent sediment
grain size characteristics: (1) wave and littoral current dominated,
characterized by relatively high energy conditions and relatively
coarse-grained, relatively well sorted sediments, (2) wave dominated,
characterized by moderate energy conditions and relatively coarse-
grained, relatively poorly sorted sediments, and (3) sheltered areas,
characterized by low energy conditions and fine-grained, poorly sorted
sediments. He points out that the man-made spoil banks, as well as
the underwater extension of Cedar Point act as barriers, sheltering
the western part of the bay and the area just southwest of Cedar
Point, respectively from the intense wave activity generated by

strong northwest winds. Throughout the bay, the turbulent forces
resulting from wave activity are probably the major factor controlling the distribution of sediments once they have entered the bay.

The surface sediments offshore from Cedar Point are more well
sorted than those in the southeastern and southcentral parts of the
bay. The underwater extension of Cedar Point spit receives the direct brunt of wave activity from the open lake and tends to partially
shelter the southeastern part of the bay. Wave activity generated
by strong winds from the north and east quadrants would tend to
keep fine-grained material in suspension and therefore reduce the
range of grain size in the surface sediment. This is reflected in
the fact that the sediment directly southwest of the extension of
Cedar Point is finer-grained and more poorly sorted than predicted
by the trend surface computed by McBride (1975), while those directly
northwest of the Cedar Point shoreline are coarser-grained and more
well sorted than predicted.

The primary driving forces that produce current in the Maumee
River estuary are wind tides, seiches and river discharge. The
estuary and harbor area of Maumee Bay are not greatly affected by
longshore currents because of the sheltering effect of man-made fills
(Miller, 1968). The outer parts of the bay, in the vicinity of Cedar
Point spit and North Cape, are more strongly effected by longshore
currents. Wind tides are a direct result of wind stress which
pushes water toward the leeward shore, increasing the water level
at that shore while it is depressed on the windward shore. As the
wind force diminishes, the stress cannot maintain the gradient,
resulting in a free oscillation of the lake surface or seiche.
The period for a longitudinal seiche (NE - SW) on Lake Erie is
approximately 14 hours (Verber, 1960). Wind-produced fluctuation
occurring in conjunction with prevailing low or high water have
resulted in water levels ranging from 7.5 feet below (U.S. Army,
Corps of Engineers, 1945) to 7.4 above (Carter, 1973) LWD.

The Maumee Bay shore is exposed to storm waves mainly from the
east to northeast to north. The maximum fetch distance for the
Maumee Bay shoreline is approximately 50 miles which restricts the
development of large waves. The shallow nature of the bay causes
"deep water" open lake waves to break, reform, and break again
several times before they reach the shore, thus dissipating much of
their energy (Benson, 1975). The maximum annual "deep water" wave
height which could be developed in the western basin of Lake Erie
had been calculated by the U.S. Army, Corps of Engineers (1953) to
be approximately 8.1 feet at Monroe, Michigan during the ice free
period of the year. The depth of water at which a wave breaks is
approximately 1.3 times the wave height (U.S. Army, Corps of Engineers, 1961). No detailed analysis of wave characteristics is

available for Maumee Bay but Benson (1975) stated the generalization that wave heights are lower in the bay than for the open lake due to: (1) predominately offshore winds which do not generate large nearshore waves in the bay, (2) low fetch distances when compared to other portions of the Lake Erie shoreline and (3) shallowness of the bay which precludes the formation or translation of large waves. In particular he concluded that the spoil islands adjacent to the navigation channel exert a "tremendous influence" on the wave characteristics of the bay. Waves crossing the spoil mounds interact with the bottom and break, thus acting as an offshore breakwater offering protection to the west shore of Maumee Bay when waves are from the east or northeast and for the south shore when waves are from the north or northwest. Benson also stated that the subaquaeous portion of the Cedar Point spit can influence wave activity within the bay by buffering large open lake waves from the north and northeast.

Waves approaching the shore obliquely generate littoral currents which flow parallel to the shoreline and may attain velocities capable of eroding and transporting particles as large as sand and gravel, particularly during storm periods. The material moved and redeposited by these currents, generally sand and gravel, is termed littoral or longshore drift. There is little or no littoral drift within Maumee Bay, but its low, clay banks are experiencing a shore recession rate up to 20 feet per year (Herdendorf, 1975) with an average of 5 feet per year or volumetrically, 1.2 cu. yds. per foot per year (Benson, 1975). The original U.S. Land Survey of the south shore of Maumee Bay in 1834 shows the shoreline to be between 1000 to 1400 feet lakeward of the present shore (U.S. Army, Corps of Enginnners, 1961). However, littoral drift moving south along the Michigan shore is responsible for the formation of Woodtick Peninsula. Similarly, drift moving predominately northwest along the 15-mile stretch of Ohio shores of Lake Erie between Locust Point and Cedar Point has formed the Cedar Point - Turtle Island spit.

Water Quality

The Maumee River estuary, for much of its 15-mile length is polluted. Despite its "gross" pollution designation, few of the numerical water quality standards are violated (Horowitz, *et al*, 1975). Dissolved oxygen and fecal coliform bacteria, particularly downstream from the Anthony Wayne Bridge, are the principal parameters which exceed the standards. Over one million tons of sediment flow down the river to the estuary annually. This produces continually turbid conditions in the lower reaches of the river. Also carried by the river are substantial quantities of fertilizers and pesticides. Approximately 160,000 tons of nitrogen, phosphorus and potassium and 16,000 tons of herbicides, fungicides, and insecticides are applied

to farm lands within the basin annually. Turbidity in Maumee Bay
decreases 130 percent from the river mouth to navigation light No.
30 (five miles offshore) and reaches background concentrations
approximately 15 miles into Lake Erie (Pinsak and Meyer, 1975).
The general trend is for high concentrations of nutrients, chloride,
silica, calcium, sodium, magnesium and potassium in the river to
decrease northeastwardly across the bay. The Toledo Lucas County
Port Authority (Fraleigh, et al., 1975) conducted a comprehensive
investigation in 1974 of water quality and biota of Maumee Bay with
particular emphasis on the proposed diked disposal area adjacent to
Harbor View. This study found that Lake Erie has a pronounced effect
on the water quality of the bay; the dilution effect of the lake in
summer tends to improve water quality in the bay. In general, good
quality lake water enters the bay from the east and poor quality
water enters from the south and west via the Maumee and Ottawa Rivers.

METHODS

In February 1975, the NASA Lewis Research Center (LeRC) instal-
led a radio facsimile receiver aboard The Ohio State University,
Center for Lake Erie Area Research (CLEAR) vessel R/V Hydra (65-ft)
at South Bass Island in western Lake Erie. LeRC then commenced
transmitting to the R/V Hydra images of Maumee Bay ice conditions
which were obtained by aircraft survey as part of the Great Lakes
Winter Navigation Program.

Commencing in early March, on clear days when there was no significant ice cover, LeRC collected large area multispectral scanner
imagery of the entire western basin of Lake Erie from a altitude of
10,000 feet. The coverage was adequate to monitor the dispersion
of the sediment plume from the Maumee River and its interaction with
the Detroit River water mass. Based on Maumee River flow rates
obtained from the USGS gauge at Waterville, Ohio and the dispersion
patterns shown by the scanner imagery, a cruise schedule for CLEAR
research vessels was formulated for mid-March.

On clear days prior to the scheduled cruise additional large-
area imagery were obtained for final ship planning and determining
dispersion patterns. On most of the days of the cruise LeRC overflew
the CLEAR vessels and collected imagery to assist in directing these
vessels to important sampling stations and for correlation with the
shipboard in situ water samples. The altitude and actual flight
lines for the over-flights depended upon weather conditions. At a
flight altitude of 5,000 feet the area west of a line between Catawba
Island, Ohio and Stony Point, Michigan could be covered; whereas at
2,500 feet the coverage was only west of a line from Locust Point,
Ohio to Stony Point. The imagery in all cases were radioed to the
R/V Hydra as soon as feasible after acquisition.

The R/V Hydra and a 22-ft Boston Whaler were used to conduct the 7-day cruise. These boats were scheduled to leave Put-in-Bay at the time of peak run-off and visit 10 stations between South Bass Island and the Maumee River based on the aircraft imagery. However, the peak run-off of the Maumee River occurred in late February, at a time when boats were iced-in at Put-in-Bay Harbor. The boats were able to break through the ice on 18 March and the survey was initiated on that date. During the next 3 days the boats operated in the Maumee River/Bay area visiting approximately 20 stations each day based on aircraft imagery (Figure 2). On the next three days, operations were hampered by rough lake conditions, but an additional 13 stations were visited by the R/V Hydra before returning to Put-in-Bay.

Surface and bottom water samples were taken at most stations with a submersible pump system or a 5-liter Niskin sampler. Filtered samples (Whatman GF/C filter pads) were analyzed for soluble reactive phosphorus, ammonia, nitrate + nitrite and dissolved silica and unfiltered samples for total phosphorus on board ship with a Technicon Auto Analyzer II. Alkalinity was determined at selected stations by HCL-titration. Calcium, magnesium, chloride and sulfate were measured with Orion, specific-ion probes fitted to an Orion model 407 pH meter. Cadmium, chromium, copper and zinc were determined using standard atomic absorption procedures prescribed for a Perkin-Elmer model 303AA. Mercury was analyzed using cold vapor flameless atomic absorption procedures with a PE-303 AA. Determinations of particulate carbon and nitrogen were performed according to standard procedures for a Perkin-Elmer model 240 elemental analyzer. Chlorophyll was determined with a Varian spectrophotometer, programmed for a, b, and c wave lengths and linked to a computer for calculating pheopigment corrections. Chlorophyll was analyzed according to procedures outlined by Strickland and Parsons (1968). Concentrations of chlorophyll were calculated using equations of SCOR/UNESCO (1966). Pheopigments were determined according to Lorenzen (1967). A Martek Mark II model A monitor system (an integrated multi-parameter instrument for simultaneous measurements) was utilized for measures of temperature, conductivity, pH, depth, and dissolved oxygen. Transparency was measured with a 30-cm Secchi disk, transmissibility with a Martek transmissometer, light penetration with Kahl submarine photometer and turbidity with a Hach model 2100 turbidometer. Suspended solids were determined by Eco-Labs, Inc., using standard muffle furnace techniques. Phytoplankton samples were taken with a Niskin sampler, zooplankton samples with a 50-cm diameter, no. 20 net (80 micron mesh) and benthic invertebrates with a Ponar sampler using a no. 40 sieve (0.425 mm mesh).

Phytoplankton samples were taken one meter below the surface. The samples were preserved with Lugol's solution and sedimented in 10 ml settling chambers (Utermohl, 1931) for examination with a

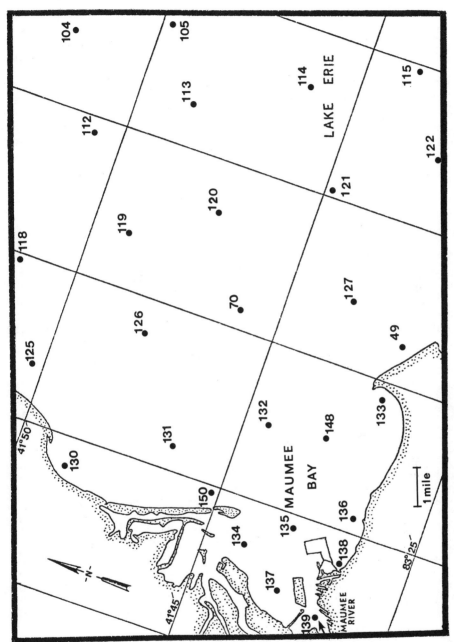

Figure 2. Station location map for Maumee Bay and western Lake Erie.

Leitz Divert inverted microscope equipped with phase contrast illu-
mination. Counts were made at 200x magnification; 50 random fields
were examined. Calculations were performed to express phytoplankton
counts as cells per milliliter. Zooplankton samples were collected
by vertical tow, bottom to surface. Before preservation, zooplankters were relaxed with club soda and preserved with 5% buffered for-
malin. Samples were counted using a stratified counting procedure.
Approximately 200 adult crustaceans and 250-300 immature crustaceans
and rotifers were enumerated. Counts and identifications were done
using a Wild M5 binocular dissecting microscope at 50x in conjunction with a Ward Zooplankton counting wheel. Numbers of zooplankters
were calculated for each tow and expressed as number of plankters per
cubic meter.

RESULTS AND DISCUSSION

Temperature

    Figure 3 illustrates the composite surface temperature of the
lower Maumee River, Maumee Bay and the adjacent part of western Lake
Erie during the March cruise. Three zones of high temperatures were
observed: (1) Maumee River mouth, (2) discharge from Toledo Edison
Company's Bayshore Power Station immediately east of the river mouth
and (3) discharge from Consumers Power Company generating plant at
the base of Woodtick Peninsula. The temperature of Maumee Bay decreased progressively from the river mouth ($6^{o}C$) to station 70 at the
Toledo Harbor Lighthouse ($3^{o}C$). Northeast of the lighthouse a wedge
of colder ($1^{o}C$) Detroit River water appears to have been moving toward the bay. Water lakeward to the 3° contour (seven miles) appears
to be directly related to Maumee River outflow.

Conductivity

    Conductivity contours (Figure 4) show the Maumee River as a
source of highly mineralized water flowing into Maumee Bay. The conductivity patterns are similar to those observed for temperature,
except that the river seems to be the only source of highly conduc-
tive water. A definite flow of Maumee River water toward the east
is indicated by a lobe in the contours projecting in that direction
beyond the bay. Based on the decrease in conductivity across the bay
(600 to 300 umhos/cm), the concentration of dissolved ions at the
lighthouse is about half of that found at the river mouth indicating
a rapid dispersion rate. A wedge of low conductance (200 umhos/cm)
Detroit River water was noted moving toward the southwest.

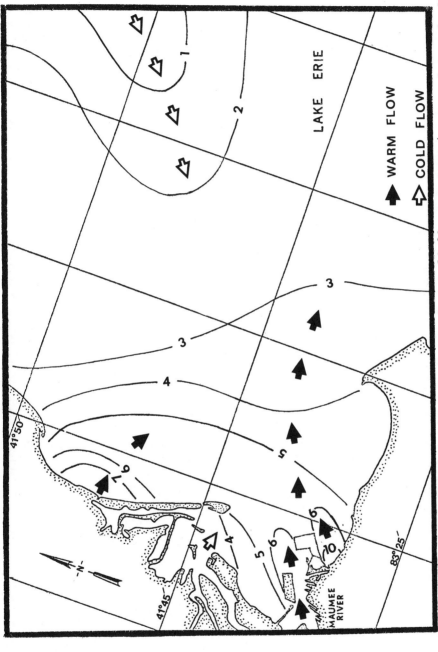

Figure 3. Temperature of surface water in Maumee Bay on 19-20 March 1975 (contour interval: 1°C).

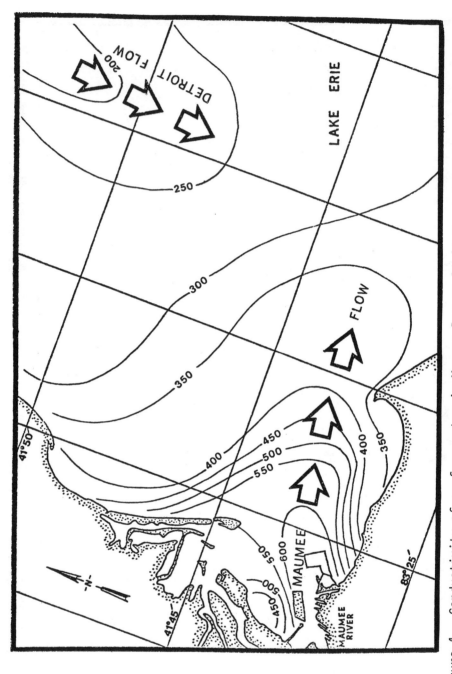

Figure 4.   Conductivity of surface water in Maumee Bay on 19-20 March 1975 (contour interval: 50 μmhos/cm).

Suspended Particles

The area of greatest turbidity was found along the south and
east shores of Maumee Bay and along the lake shoreline southeast of
Cedar Point at the eastern terminus of the bay (Figure 5).   Here
turbidity exceeded 100 Jackson Turbidity Units (JTU), twice the
values at the river mouth.   Thus, the river was not the dominant
factor influencing turbidity in the eastern part of the bay.   Wave
action appears to have resuspended bottom sediments causing high
values of turbidity which extended from Cedar Point to the navigation channel.   The extent of resuspended material is well-illus-
trated by the NASA multispectral scanner imagery for 23 March 1975
(Figure 6).   Figure 7 is another scanner imagery taken one year later
(25 March 1976) and demonstrates the persistence of the turbidity
patterns and their extent throughout the western basin of Lake Erie.
Enlarged LANDSAT-1 imagery for 28 May 1975 (Figure 8) depicts a turbidity plume created by the passage of a large vessel in the Toledo
navigation channel.   Plumes such as this have been observed 20 miles
lakeward of the river mouth.

The maximum concentration of suspended solids (80 mg/1 dry
weight) was found between Cedar Point and North Cape, where Maumee
Bay joins western Lake Erie.   Values of only 50mg/1 which correlated
well with the patterns of turbidity were measured at tne mouth of
the Maumee River.   Measurements 20 miles northeast of the river mouth
yielded values less than 10 mg/1 in Detroit River flow (5 JTU).
Water transparency showed similar patterns with inverse values.   Low
transparency (0.25 meters) in the river and along Cedar Point indicated again the importance of both run-off and wind and wave action in
controlling water clarity.   Transparency did not increase above 0.5
meter within 15 miles of the river mouth.   Beyond that point the
clarity improved rapidly to 2.0 meters at the northeastern portion
of the study area.

Nutrients

Nitrate + nitrite nitrogen and ammonia (Figure 9) demonstrated
again the significant effect of the Maumee River on Maumee Bay.
River mouth values were approximately three times higher than station 70 (lighthouse).   Soluble reactive phosphorus showed a typical
dispersion pattern, high at the river stations and falling as the
flow progressed through the bay (Figure 10).   The highest river
value was 126 ppb and the lowest bay value was 18 ppb.   All river
measurements for silica were above 6000 ppb.   This level was maintained lakeward to North Cape and Cedar Point where the concentration decreased very gradually as the flow entered western Lake Erie.

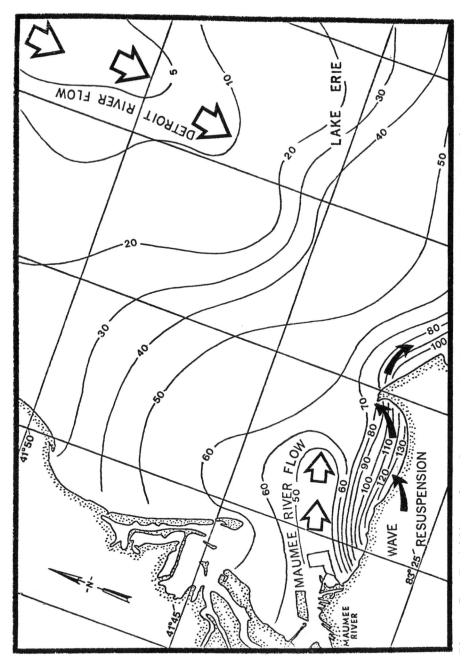

Figure 5.   Turbidity of surface water in Maumee Bay on 19-20 March 1975 (contour interval: 5-10 JTU).

Figure 6. Multispectral scanner image of Maumee Bay and south shore of western Lake Erie on 23 March 1975.

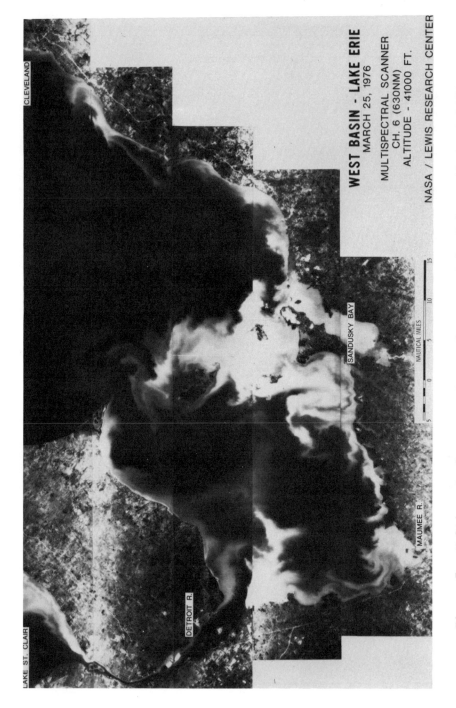

Figure 7. Multispectral scanner image of western Lake Erie on 25 March 1976.

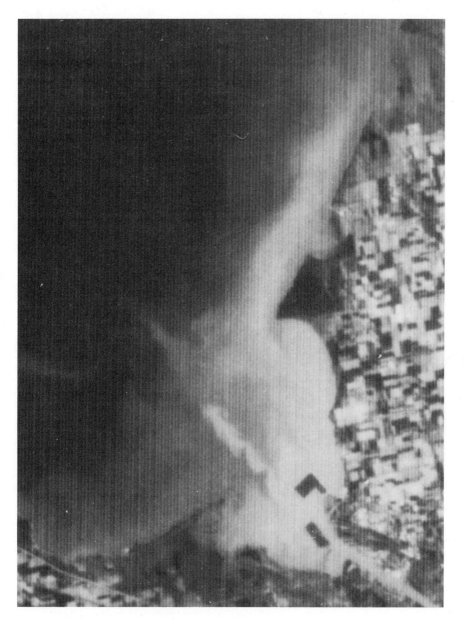

Figure 8. LANDSAT-1 image (band 5) of Maumee Bay on 28 May 1975, showing a sediment plume initiated by the passage of a large cargo vessel.

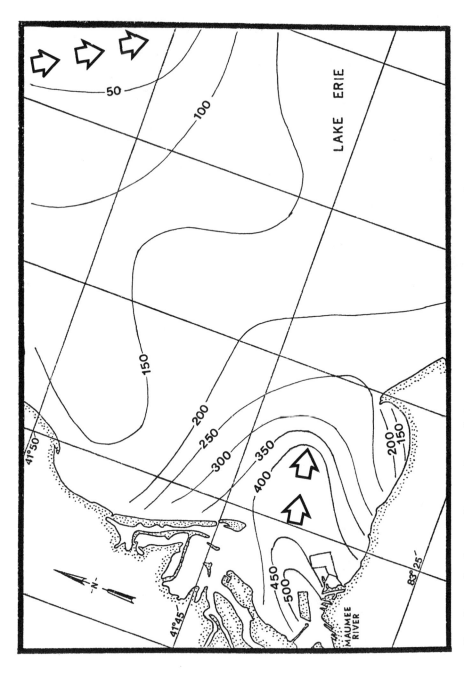

Figure 9.   Ammonia in surface water of Maumee Bay on 19-20 March 1975 (contour interval: 50 ppb).

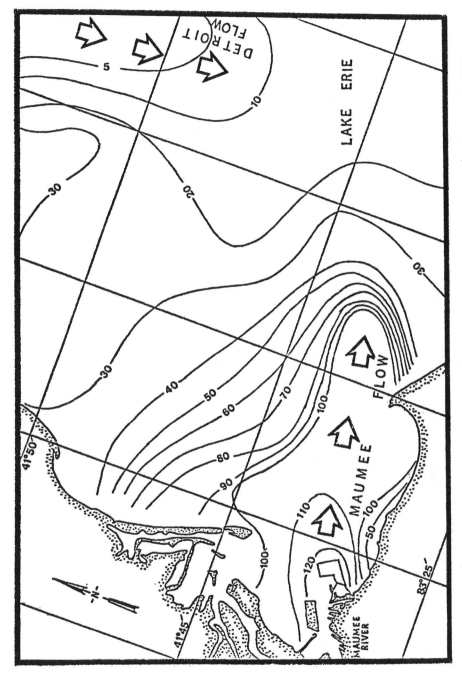

Figure 10. Soluble reactive phosphorus in surface water of Maumee Bay on 19–20 March 1975 (contour interval: 10 ppb).

Chlorophyll

Chlorophyll a concentrations indicate patterns of phytoplankton density (Figure 11). Low concentrations were found near the mouth of the Maumee River and the inner portion of Maumee Bay. Low phytoplankton populations in these areas were probably the result of inhibition due to high turbidity. The highest concentrations were found in a zone from the lighthouse where Maumee Bay joins western Lake Erie to the southeast along the Ohio shoreline (Figure 11). This zone was also rich in nutrients. Low chlorophyll concentrations at the northeastern limits of the study were probably the result of low temperature and nutrient levels. High suspended solids were observed in the zone of high chlorophyll values. High ash weight values (volatile solids) in this zone indicated that the suspended solids contained a high percentage of organic material.

Phytoplankton

Phytoplankton and zooplankton samples were collected in a line from the Maumee River mouth into the western basin of Lake Erie on 20 March. The line of stations monitored, 141, 140, 135, 132, 70, 120, 113, 105 and 101, was located in the navigation channel extending from the Toledo harbor northeast 21 miles into the western basin. (Figure 2). Station 141 and 140 (3 and 6 miles upstream from the mouth respectively) were representative of lower river conditions. The remaining stations leading to 101 (3 miles northeast of station 105) demonstrated a gradient from harbor to open lake conditions. Station 101 typified the open western basin which was strongly influenced by the Detroit River carrying upper Great Lakes water.

The phytoplankton community changed significantly both quantitatively and qualitatively from the harbor to the open basin (Figure 12 and Table 1). During the spring the western basin is characteristically very turbid. At all stations sampled, the quantity of seston in the water was high with the inorganic fraction consistently greater than the organic fraction. A sharp increase in the quantity of suspended material was detected at Station 70. Shoreward from station 132, only 2 ml of whole water could be settled for phytoplankton analysis due to the large quantities of seston which covered the settled plankters. (Normally, 10 ml was settled.) At all stations a significant portion of the detrital material was algal in origin. Numerous cells in various degrees of decomposition were identificable. *Pediastrum simplex* and numerous taxa of diatoms were observed frequently. At station 101 *Fragilaria crotonensis* was found in concentrations up to 144 frustules/ml. All of these cells were in various stages of decompostion.

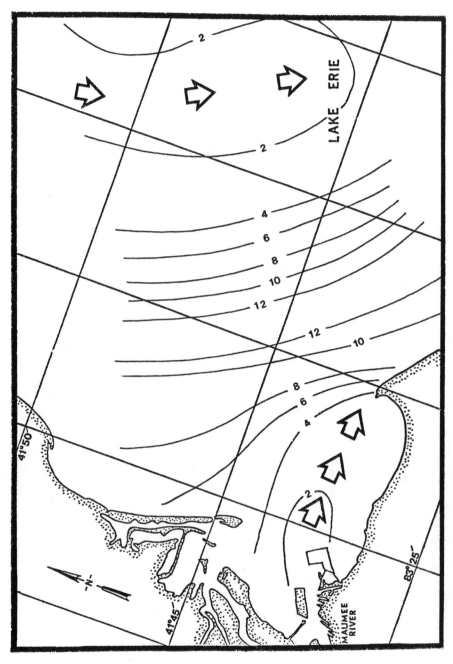

Figure 11. Chlorophyll a (corrected) in surface water of Maumee Bay on 19-20 March 1975 (contour interval: 2 µg/l).

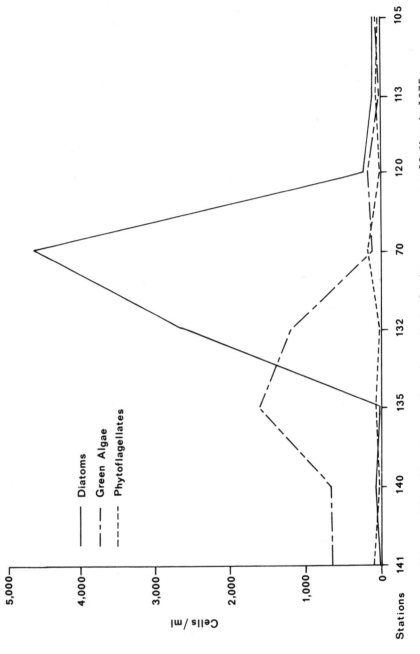

Figure 12.   Abundance of major phytoplankton groups on 20 March 1975.

TABLE 1

PHYTOPLANKTON POPULATIONS ON 20 MARCH 1975
Cells per Milliliter

Taxa	141	140	135	132	070	120	113	101
Diatoms								
Fragilaria crotonensis				61	107	14	214	144
Cyclotella sp.		37	37	37		5		5
Navicula sp.								5
Melosira granulata								2
Asterionella formosa				12	156	17	24	34
Tabellaria fenestrata					19	24	41	14
Stephanodiscus astraea					5	29	34	
Stephanodiscus hantzschii				2545	3342	158	2	
Gomphonema sp.							2	
Coscinodiscus rothii						2		
Green Algae								
Ankistrodesmus falcatus var. *mirabilis*			12	12	24	19	85	19
Oocystis Borgei								2
Scenedesmus quadricauda			24		5			
Kirchneriella lunaris						2	12	
Pediastrum duplex					1			
Pediastrum simplex								
Schroederia setigera		1	12					
Crucigenia tetrapedia	97							
Phytoflagellates								
Dinobryon divergens					29		17	24
Rhodomonas minuta			61	122	15		2	19
Cryptomonas erosa						5	5	5
flagellates spp.	657	670	1158	1072	78	172	43	17
Blue-Green Algae								
Oscillatoria sp.	1							

Three groups dominated the phytoplankton commmunity at this time; diatoms, green algae and phytoflagellates. The dominant group changed from the open basin station 101 to station 141 (Figure 13). Only in the outer portion of the study area, stations 101-113, were the *Chlorophyta* significant with *Scenedesmus quadrigula* and *Ankistrodesmus falcatus* being the major taxa. Diatoms comprised the dominant group from stations 132 to 101 with *Asterionella formosa* and *Tabellaria fenestrata* predominating. At stations 120 through 132 a small centric diatom (8μ) believed to be c.f. *Stephanodiscus hantzschii* was found in high concentrations (3,000 cell/ml). This population peaked in the region of station 70 and diminished shoreward. Station 135 had only a small diatom population and the inner channel and river stations also contained low populations. The third group, phytoflagellates, was comprised of Crysomonadinae and Cryptomonadinae. The taxa primarily representing this group were *Rhodomonas minuta, Cryptomonas erosa,* and *Crysococcus* sp. Due to the heavy sediment content, many of these small organisms were difficult to identify. The phytoflagellates were the dominant group in the Maumee Bay stations.

Only at stations 70 and 132 (and to a lesser degree 135) were there substantial populations of phytoplankton. These denser localized populations are attributed to c.f. *Stephanodiscus hantzschii* which reached a maximum of 3,300 cells/ml at station 70. Stations both toward the open basin and shoreward contained low numbers of phytoplankters.

Chlorophyll a corrected and uncorrected for pheopigments, was analyzed for each of the stations where phytoplankton was collected (Figure 11). Generally the pattern for chlorophyll a was similar to that of the phytoplankton. The outer stations, 101-120, had the lowest chlorophyll a values. An increase was evident at station 120 with the peak concentration occurring at station 70 (12 μg/1). The cholorophyll concentration decreased from station 70 and became more stable, nearly 3μg/1. shoreward from station 135. Pheopigment distribution did not show the definite pattern demonstrated by chlorophyll a. Pheopigments comprised from 20 percent to 40 percent of the total chlorophyll measured. This high percentage appears to be characteristic of the western basin during periods of high sediment resuspension.

The chlorophyll peak occurring at station 70 coincided with the populations of *Stephanodiscus hantzschii* found in this region. Shoreward from station 135 chlorophyll remained constant, although the phytoplankton occurred in low densities. This distortion may be explained by the resuspension of decomposing cells containing chlorophyll as yet not totally degraded. (Vallentyne, 1955).

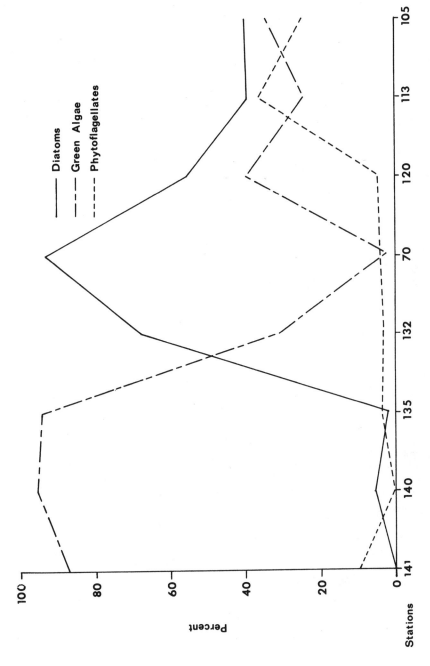

Figure 13. Percent composition of major phytoplankton groups on 20 March 1975.

During the summer months the region near station 70 is one of the most productive in the western basin. Chlorophyll a values average approximately 25 µg/l throughout the summer. During March 1975 the chlorophyll values were considerably lower due to low temperature and high turbidity.

Zooplankton

The zooplankton community also demonstrated changes in populations from the river mouth to the open lake (Figure 14 and Table 2). Rotifers, cladocerans, and copepods were identified along the series of stations within the navigation channel. Only the rotifers and copepods were important to the zooplankton community. Cladocerans did not comprise more than 2 percent of the total community at any station. The cladoceran population in Lake Erie does not normally peak until mid-summer. The cladoceran species encountered were *Bosmina longirostris*, *Eubosmina coregoni*, *Chydorus sphaericus*, *Daphnia galeata mendotae* and *Leptodora kindtii*. In general the zooplankton had the same basic distribution as did the phytoplankton. The river stations and lakeward to station 135 contained very low concentrations of plankters. From station 132 to 120 the number of plankters increased greatly. A decrease was evident from station 120 to the outer open lake stations.

Rotifer populations were quite important to the total zooplankton community comprising from 20 percent to 67 percent of the total plankters (Figure 15). Shoreward from station 70, rotifers were nearly 50 percent of the total zooplankton. Lakeward to station 101 they comprised 20 percent to 40 percent of the total community. The dominant rotifers present were *Synchaeta* spp. and *Notholca* spp. *Synchaeta spp.* occurred in the highest relative abundance at all stations with *Notholca* generally second. *Polyarthra* and *Keratella* spp. were present in relatively low concentrations. The most taxonomically diverse area was in the lower river. The rotifers found exclusively in this region were *Euchlanis*, *Conochiloides*, *Pomphylx*, *Rotaria* and *Conochilus*. Several of these taxa are known indicators of eutrophication.

The copepods represented the other important segment of the zooplankton community. During this time of year the immature forms, nauplii and copepodids are predominant (Figure 15). They composed 90 percent of the total copepod population from the river to station 135 and about 60 percent lakeward from 135. The adult copepods did not become important until station 70. Both the cyclopoid and calanoid copepods showed similar distribution patterns. Both were present in low concentration inshore to station 70. Lakeward of station 70, individual taxa increased depending on location. *Cyclops bicuspidatus thomasi* was the only cyclopoid copepod of importance

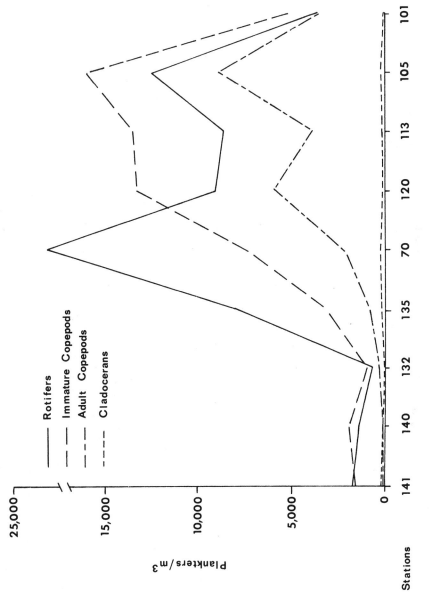

Figure 14. Abundance of major zooplankton groups on 20 March 1975.

TABLE 2

ZOOPLANKTON POPULATIONS ON 20 MARCH 1976
Plankters per Cubic Meter

Taxa \ Stations	141	140	135	132	70	120	113	105	101
ROTIFERS									
Asplanchna priodonta	63	20				76			
Brachionus angularis	42	60	41						
B. calyciflorus	125			129	154				
B. urceolaris		20		129					
Chromogaster ovalis			21						
Conochilis unicornis			10						
Filinia terminalis	83	60		258					48
Kellicottia longispina				258	154	153	134	420	
Keratella cochlearis	21	60	21		770	1,221	671	560	48
K. crassa	125							140	
K. quadrata	188	240	31		1,848	229	268		96
Notholca spp.	42	80		516	1,694	763	1,476	979	144
N. foliacea		20							
N. squamula		40		1,290	770	534	537	1,119	672
N. strata	42	100	21	129		382	134	280	
Polyarthra spp.	63	120			308	458			528
Rotaria sp.			83						
Synchaeta spp.	771	520	311	4,387	15,708	5,340	5,234	8,814	2,160
Rotifer spp.	84	140	21	903					
Total Rotifers	1,646	1,480	662	8,000	22,946	9,080	8,454	12,311	3,696
CLADOCERANS									
Bosmina longirostris	42	20	21	39	35				
Chydorus sphaericus			10		23	51			
Daphnia galeata				6	23		17		
Eubosmina coregoni					23	25		140	
Total Cladocerans	42	20	31	45	104	76	17	140	0
COPEPODS									
Nauplii	1,562	1,920	880	2,710	6,468	9,766	13,071	12,871	4,656
Calanoid copepodids							134		96
Cyclopoid copepodids	42	40	83	387	1,078	3,434	268	3,078	432
Harpactacoid copepodids									
Canthocamptus			31		23				
C. bicuspidatus thomasi			10	342	244	1,829	151	1,260	619
Cyclops vernalis	21			6					
Diaptomus ashlandii				200	812	2,642	1,243	4,970	1,757
D. minutus				58	371	784	890	1,050	504
D. oregonensis				19	12	127	235	280	115
D. sicilis				123		508	1,327	1,190	691
Eucyclops auilis				19					14
Limnocalanus macrurus				6		25	50	175	58
Paracyclops fimbriatus			10						
Tropocyclops prasinus					12	25	17		
Total Immatures	1,604	1,960	963	3,097	7,546	13,200	13,420	15,949	5,184
Total Adults	21		52	774	2,076	5,944	3,914	8,925	3,758
Total Copepods	1,625	1,960	1,014	3,871	9,622	19,144	17,334	24,874	8,942
TOTAL	3,312	3,460	1,708	11,916	32,672	28,300	25,805	37,325	12,638

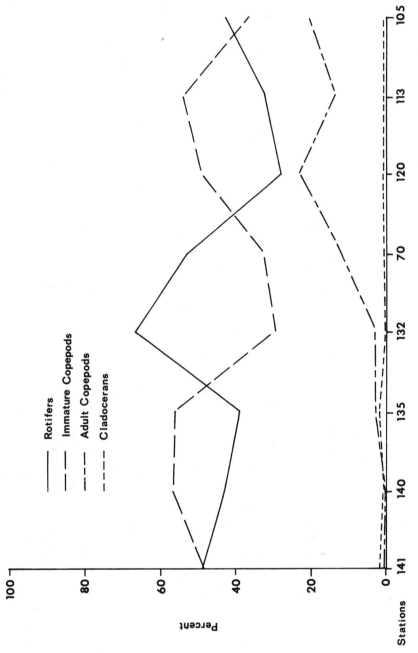

Figure 15. Percent composition of major zooplankton groups on 20 March 1975.

and was most abundant at station 120. Of the calanoid copepods, *Diaptomus ashlandii*, *D. sicilis* and *D. minutus* were dominant. All three had similar patterns of distribution with the largest populations occurring at the most lakeward stations. *D. ashlandii* was the most prevalent species.

CONCLUSIONS

The same general distribution pattern for phytoplankton and zooplankton was observed. Low concentrations of organisms were found in the Maumee River and Maumee Bay near the river mouth. Populations of plankters peaked in the area of stations 70 and 120. Low concentrations were again found at the northeast limit of the study area (stations 101 and 105).

No one factor could be determined as inhibitory to organisms in and near the Maumee River. A considerable difference in the concentration of the various parameters measured existed between stations 139 and 70. Nearly all chemical parameters measured were reduced in concentration from stations 139 to 70. The only parameter not changing significantly was transparency, while turbidity increased slightly.

The dilution of the chemically rich river water as it enters the lake is affected by many factors. During the spring run-off period, flows are high. A major portion of the river flow remains within the navigation channel for some distance offshore before dispersing. At stations 135 and 132 the river water was greatly influenced by the lake. By station 70 the concentration of most water quality parameters had been reduced by 50 percent (Table 3). The river flow entering the lake is greatly altered by winds. Southwest or west winds will carry the river plume into the western basin while northeast winds tend to push western basin water up into the river mouth. On 20 March wind conditions were relatively calm, thus river flow into the Maumee River was not influenced by this factor.

The concentration of the two major nutrients, nitrogen and phosphorus, were greatly reduced in transit across Maumee Bay. Total phosphorus was reduced 41 percent at station 70. The soluble reactive phosphorus was 91 percent of the total phosphorus at station 140 but at station 70 it comprised only 64 percent. This indicates a source of the soluble form near the river mouth and resuspension of sedimented insoluble phosphorus offshore. Phosphorus was never found at concentrations that would be considered limiting to algal growth. Total inorganic nitrogen was reduced 56 percent from station 140 to station 70. Although the NO_3 and NO_2 fraction was reduced 55 percent the concentration was very high at station 70. NH_3 concentrations in and near the river mouth were approximately 500 ppb. At

TABLE 3

GRADATION OF WATER QUALITY FROM MAUMEE RIVER TO WESTERN LAKE ERIE

Parameter	Station 140	Stations 140 to 70 % Change	Station 70	Stations 70 to 101 % Change	Station 101	Stations 140 to 101 % Change	Units
Conductivity	575	<36	370	<35	240	<58	μmohs/cm
Transparency (Secchi)	0.37	<46	0.20	>87	1.5	>76	meters
Turbidity	45	>32	66	<93	4.5	>100	JTU
Calcium	110	<50	55	<18	45	<59	mg/l
Chloride	177	<51	87	<39	53	<70	mg/l
Sulfate	102	<53	48	<62	18	<82	mg/l
Total Inorganic Nitrogen	6614	<56	3011	<86	441	<94	ppb
Nitrate + Nitrite	6195	<55	2805	<87	363	<94	ppb
Ammonia	619	<66	206	<62	78	<87	ppb
Total Phosphorus	123	<41	73	<58	31	<75	ppb
Soluble Reactive Phosphorus	113	<58	47	<61	18	<84	ppb
Dissolved Silica	6320	<11	5650	<97	155	<97	ppb
Corrected Chlorophyll a	3.34	>73	12.08	<96	0.46	<86	μg/l

this concentration NH_3 has been reported to be inhibitory to photosynthesis (Abeliovich, 1976). At station 70 the concentration was reduced to about 200 ppb which is below reported toxic levels. Neither turbidity nor transparency had improved at this station due to the heavy sediment load carried by the river and the great quantities of resuspended bottom materials in the shallow waters of the bay.

The low concentrations of phytoplankton and zooplankton in the Maumee River and shore area of the channel are attributed to high flow and turbidity, extremely high concentrations of chemicals, especially NH_3, and possible other toxic substances entering the lake from the river. Of the trace metals tested in the river and bay waters only zinc and copper exceeded Ohio EPA standards. Station 70 appeared to be located at the point in the bay of optimum growth conditions for mid-March. Dilution of the major chemicals entering the bay had occurred but the temperature of the water was still warm from river flow.

The number of planktonic organisms again decreased lakeward from station 70. The dilution effect continued to station 101 with many of the parameters being reduced by over 75 percent from the river mouth. Neither nitrogen or phosphorus were found in limiting concentrations, even at the most lakeward station. Turbidity was much improved (100 percent) in this area. The major parameter which may have biological importance was temperature. The temperature in the open western basin was only $1°C$ in mid-March and this may explain the low phytoplankton and zooplankton activity in the region of the western basin influenced by Detroit River flow.

ACKNOWLEDGEMENTS

The authors wish to acknowledge the assistance of the National Aeronautics and Space Administration, Lewis Research Center in providing multispectral scanner imagery of western Lake Erie. Dr. Richard Gedney and his staff were extremely helpful in planning and executing the field statistics. This project was conducted in support of Task D of the Pollution from Land Use Activities Reference Group of the International Joint Commission. Support of this study was also obtained from the U.S. Environmental Protection Agency, Large Lakes Field Station, at Grosse Ile, Michigan. The authors gratefully acknowledge the help of Dr. John E. Zapotosky and his staff aboard the R/V Hydra for collecting the field data.

REFERENCES

Abeliovich, A. and Yosef Azov. 1976. Toxicity of ammonia to algae in sewage oxidation ponds. Applied and Environ. Microbiol. 31(6):801-806.

Benson, D.J. 1975. Maumee Bay erosion and sedimentaion. U.S. Army, Corps of Eng., Draft Rept. Contract DACW 35-75-C-0038. 214 p.

Brant, R.A. and C.E. Herdendorf. 1972. Delineation of Great Lakes estuaries. Proc. 15th Conf. Gt. Lakes Res., Internat. Assoc. Great Lakes Res. 1972: 710-718.

Carter, C.H. 1973. The November 1972 storm on Lake Erie. Ohio Dept. Nat. Res., Div. Geol. Surv. Infor. Circ. 39. 12 p.

Charlesworth, L.J. 1974. Maumee Bay dike impact study, phase I: initial bottom sediment sampling. U.S. Army, Corps of Eng. Contract DACW 35-75-M-0189. 309 p.

Fraleigh, P.C., J.C. Burnham, G.H. Gronau, T. Kovacik and E.J.Tramer. 1975. Maumee Bay environmental quality study. Draft Final Rept., Toledo-Lucas County Port Authority.

Great Lakes Basin Commission. 1975. Surface water hydrology. Gt. Lakes Basin Framework Study, Appendix 2, Gt. Lakes Basin Comm., Ann Arbor. 133 p.

Hartley, R.P. 1960. Sand dredging area in Lake Erie. Ohio Dept. Nat. Res., Div. Shore Erosion Tech. Rept. 5. 79 p.

Herdendorf, C.E. 1970. Sand and gravel resources of the Maumee River estuary, Toledo to Perrysburg, Ohio. Ohio Dept. Nat. Res., Div. Geol. Surv. Rept, Invest. 76. 19 p.

Herdendorf, C.E. 1975. Shoreline changes of Lakes Erie and Ontario. Bull. Buffalo Soc. Nat. Sci. 25(3): 43-76.

Horowitz, J., J.R. Adams and L.A. Bazel. 1975. Water pollution investigation: Maumee River and Toledo area. U.S. Environ. Protect. Agency EPA-90519-74-018. 170 p.

Lorenzen, C.J. 1967. Determination of chlorophyll and pheo-pigments spectrophotometric equations. Limnol. Oceanogr. 12:343-346.

McBride, R.T. 1975. Distribution of recent sediment in Maumee Bay, western Lake Erie. M.S. Thesis, Dept. Geol. Univ. of Toledo. 155 p.

Miller, G.S. 1968. Currents at Toledo Harbor. Proc. 11th Conf. Gt. Lakes Res. Internat. Assoc. Gt. Lakes Res. 1968:437-453.

Ohio Division of Water. 1960. Water inventory of the Maumee River basin. Ohio Dept. Nat. Res., Div. Water, Water Plan Inven. Rept. 11. 112 p.

Pinsak, A.P. and T.L. Meyer. 1975. Baseline reference for Maumee Bay, Maumee River basin level B study, interim report. Gt. Lakes Basin Comm. 40 p.

Reid, G.K. and R.D. Wood. 1976. Ecology of inland waters and estuaries, 2nd Ed. D. Van Nostrand Co., New York. 485 p.

SCOR/UNESCO. 1966. Monograph on oceanographic methodology. I. determination of photosynthetic pigments in sea water. United Nations, Paris. 69 p.

Strickland, J.D.H. and T.R. Parsons. 1968. A practical handbook
    of seawater analysis. Fish. Res. Bd. of Canada Bull. 167.
    Ottawa. 310 p.
U.S. Army, Corps of Engineers. 1945. Beach erosion study, Ohio
    shore line of Lake Erie from Ohio-Michigan state line to
    Marblehead, Ohio. 79th Congr., 1st Sess., U.S. House Doc. 177.
    27 p.
U.S. Army, Corps of Engineers. 1953. Wave and lake level statistics
    for Lake Erie. Beach Erosion Board Tech. Memo. 37. 14 p.
U.S. Army, Corps of Engineers. 1961. Lake Erie shore line from
    the Michigan-Ohio state line to Marblehead, Ohio, beach ero-
    sion control study. 87th Congr., 1st Sess., U.S. House Doc.
    63. 153 p.
U.S. Army, Corps of Engineers. 1974. Confined disposal facility
    for Toledo Harbor, Ohio. Final Environ. Impact Statement.
    Detroit Dist. 85 p.
Utermohl, H. 1931. Neve wege in det quantitative erfassung der
    planktons. Verh. Intern. Ver. Limnol. 5:567-596.
Vallentyne, J.R. 1955. Sedimentary chlorophyll determination as
    a paleo-botanical method. Can. J. Bot. 33: 304-313.
Verber, J.L. 1954. Maumee Bay sand survey. Ohio Dept. Nat. Res.,
    Div. Shore Erosion. 15 p.
Verber, J.L. 1960. Long and short period oscillations in Lake
    Erie. Ohio Dept. Nat. Res., Div. Shore Erosion. 80 p.
Walters, L.J. and C.E. Herdendorf. 1975. Influence of the
    Detroit and Maumee Rivers on sediment supply and dispersal
    in western Lake Erie. Abst. 18 Conf. Gt. Lakes Res., Internat.
    Assoc. Gt. Lakes Res.

THE RETURN OF AQUATIC VASCULAR PLANTS INTO THE GREAT LAKES REGION AFTER LATE-WISCONSIN GLACIATION

Stephen J. Vesper and Ronald L. Stuckey

Environmental Biology Program and Department of Botany

The Ohio State University, Columbus 43210

ABSTRACT

This paper traces the recolonization by selected aquatic vascular plants into the Great Lakes region after late-Wisconsin glaciation. The currently widespread genera, Myriophyllum, Nymphaea, Nuphar, Potamogeton, Sagittaria, and Typha were chosen because they are most frequently recorded in pollen diagrams. The movements of these genera were inferred by obtaining published pollen diagrams in which these genera were recorded. Only those records were used which were radiocarbon dated or based on inferences from other sources which had been radiocarbon dated. The total number of pollen profiles deamed useful in the literature is about 32; but, unfortunately, most of them are from the eastern half of the United States. By plotting first occurrence data of the various genera, a time and location sequence was established for each. In general, the earliest records of most of these genera are from the Carolina-Virginia region. This may represent the southernmost extension of these genera during late-Wisconsin time. These genera are well documented next in the Massachusetts region. Then their distribution appears to spread out into Nova Scotia, Quebec, and the Great Lakes region. It appears that the Appalachian Mountains were an important barrier to the recolonization of glaciated areas by aquatic vascular plants in eastern United States. Since so few pollen diagrams exist from western United States, it is difficult to define the western or southwestern contribution to recolonization. A southern migration route in the Mississippi valley may be plausible. With additional pollen records and more complete dating of pollen diagrams, a better understanding of the movements of aquatic vascular plants may be obtainable.

283

INTRODUCTION

Before Wisconsin glaciation, the present-day Great Lakes region was undoubtedly covered with vegetation much as it is today.
It would have been a region covered with forests interspersed
with ponds, rivers, and lakes. Our knowledge of the aquatic vascular plant flora in those aquatic habitats is essentially speculative in that we may presume that many of the common genera that
are in the Great Lakes region today were there then, and that like
terrestrial plants, many of them were able during glaciation to
survive somewhere outside the ice covered region. Their survival
may have been along the margin of the glacier or farther south on
the Atlantic Coastal Plain or in the warmer waters of the present-
day Mississippi valley. This paper is a first attempt to trace
the recolonization of selected aquatic vascular plants into the
Great Lakes region following late-Wisconsin glaciation from plant
remains preserved in lake sediments and reported in pollen analytical studies.

During the past 50 years in eastern North America a sizeable
amount of literature has been written on the plant remains present
in bog and lake sediments. These plant remains are usually in
the form of pollen grains, spores, seeds, or other portions of
the plants. The bog and lake deposits have been sampled in many
areas in eastern North America and have resulted in a number of
published pollen diagrams or profiles. Particularly in the past
20 years some of the plant materials have been radiocarbon dated
thus providing a time reference point for many of the pollen diagrams. These dated pollen diagrams have allowed botanists and
paleoecologists to interpret the kinds of vegetation growing in
the areas surrounding the lake or bog whence the pollen record
was preserved. Moreover, inferences as to the kind of climate,
migration routes of plants, and time sequences for many of these
phenomenon have been inferred. Because forest species are of
prime importance to most individuals plotting and interpreting
pollen profiles, little attention has been given to aquatic vascular plants, even though some pollen profiles do record certain
genera, especially in those situations where the investigator has
identified a genus of aquatic plants and recorded it in the profile. Some investigators have either not found pollen of aquatic
vascular plants or have not reported its occurrence. Therefore,
we realize that data on aquatic vascular plants in published pollen diagrams is indeed fragmentary and incomplete, and in some
cases the data may have been retained as incidental information.

METHODS

Realizing the limitations of the data as outlined above, we did, however, study as many published pollen profiles that could be located in the literature and determined from these the extent of coverage given to aquatic vascular plants. The currently widespread genera, Myriophyllum (water milfoil), Nymphaea (water lily), Nuphar (spatterdock), Potamogeton (pondweed), Sagittaria (arrowhead), and Typha (cattail) were chosen because they are the most frequently occurring ones in the pollen profiles. Only those pollen diagrams that were radiocarbon dated were used, although records of plants from earlier studies were used if these same profiles were dated at a later time. Data from 32 published papers on pollen profiles were deemed useful for our study. The pollen profiles were numbered from 1 to 32 with each one recorded on a map of eastern North America showing its geographical location (figure 1) and each one identified in the literature used section below. For convenience in determining a time sequence, the late-Wisconsin glacial period (25,000 to 8,000 B.P.) was divided into one thousand year intervals. In each pollen profile, the first record of each of the six aquatic plant genera, if recorded, was placed into one of these thousand year intervals. The derived information for each genus was plotted on a map, resulting in six plant distribution maps (figures 2-7). On the maps, each dot is at the approximate location of the site of the pollen profile and the number associated with each dot is the first occurrence in thousands of years B.P. of that genus in that particular profile (i.e., the first time the genus occurs in the pollen diagram).

RESULTS AND INTERPRETATION

The six genera considered in this study are currently common and widespread in North America. They are rooted aquatic vascular plants of essentially quiet waters in ponds and lakes, embayment marshes, or in slow flowing rivers or streams. Nymphaea, Nuphar, Sagittaria, and Typha are all emersed plants with showy flowers or inflorescences. Myriophyllum and Potamogeton are submersed plants with the flowers in many species occurring on portions of the plant above water or at the surface of the water. From the present-day North American distribution of these six genera, it would be impossible to infer any post-glacial migration routes or patterns of dispersal, similar to those that can be inferred for species of a more limited range. Examples are those restricted species occurring almost exclusively on the Atlantic Coastal Plain and then inland either continuously or as disjuncts in the Great Lakes region (Peattie, 1922), or those whose distributions are primarily in the Mississippi valley and

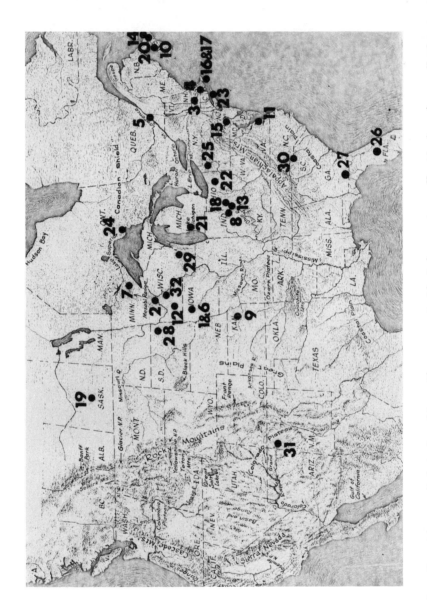

Figure 1.   Locations of dated pollen diagrams studied in eastern North America. Numbers correspond with the published references listed in the literature cited section.

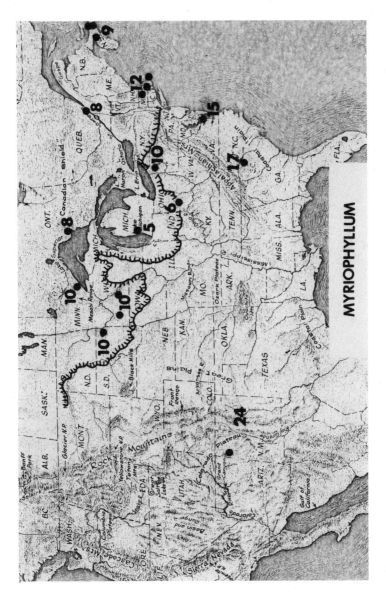

Figure 2. Locations of first occurrence records of _Myriophyllum_ (water milfoil) in pollen profiles from eastern North America. The numbers are in thousands of years B.P.

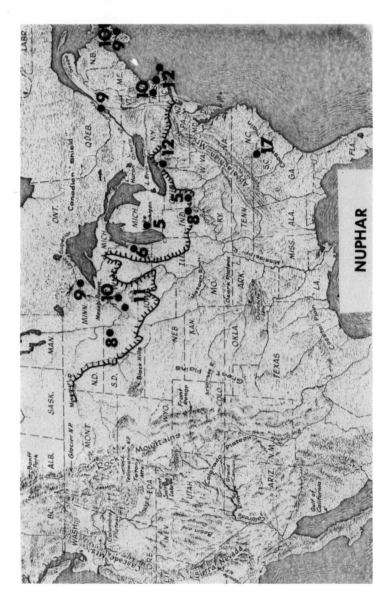

Figure 3. Locations of first occurrence records of Nuphar (spatterdock) in pollen profiles from eastern North America. The numbers are in thousands of years B.P.

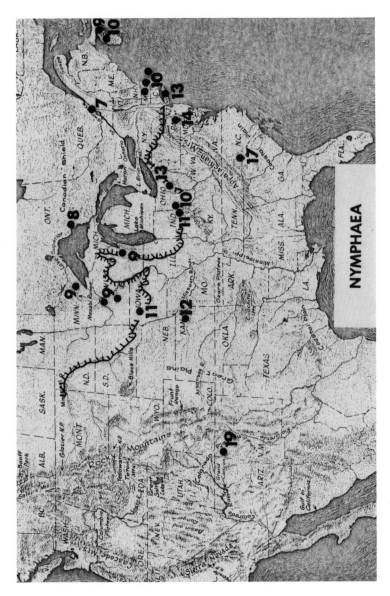

Figure 4. Locations of first occurrence records of Nymphaea (water lily) in pollen profiles from eastern North America. The numbers are in thousands of years B.P.

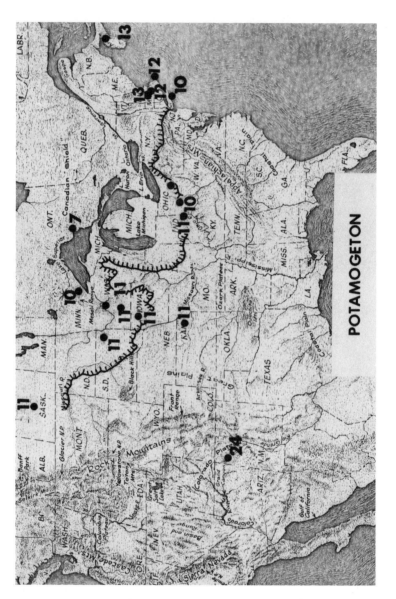

Figure 5. Locations of first occurrence records of Potamogeton (pondweed) in pollen profiles from eastern North America. The numbers are in thousands of years B.P.

Figure 6. Locations of first occurrence records of Sagittaria (arrowhead) in pollen profiles from eastern North America. The numbers are in thousands of years B.P.

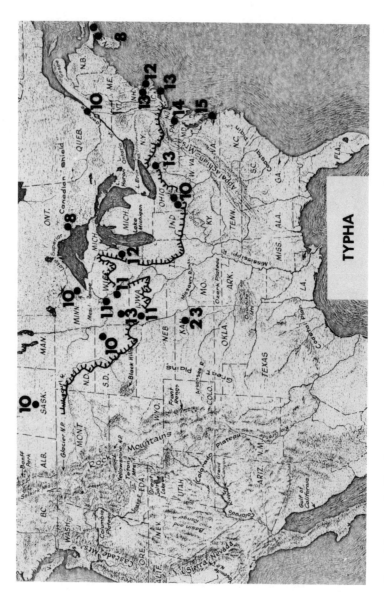

Figure 7. Locations of first occurrence records of Typha (cattail) in pollen profiles from eastern
North America. The numbers are in thousands of years B.P.

reach the northern or northeastern extension of their range in
the southern Great Lakes region, e.g., Lake Erie or the southern
portion of Lake Michigan (Gleason, 1923).

The pollen profiles reveal that Myriophyllum, Nymphaea, Nuphar, and Sagittaria are documented south of the maximum extent
of Wisconsin glaciation in the Carolina region at about 17,000
B.P. The next early occurrences for these genera are northward
along the Atlantic coast at 15,000 B.P. for Myriophyllum in southeastern Virginia, at 14,000 B.P. for Nymphaea in southeastern
Pennsylvania, and between 13 to 10,000 B.P. for these four genera
in southern New England. Typha is recorded as early as 15,000
B.P. in Virginia. These five genera therefore were disturbed on
the present-day Atlantic Coastal Plain during Wisconsin glaciation. Potamogeton appears to have no preserved history or recorded occurrences outside the glaciated area on the Atlantic
coast. However, it is possible that Potamogeton could have survived glaciation on the Atlantic Coastal Plain in areas now covered by the Atlantic Ocean. In the glaciated area, Myriophyllum,
Nymphaea, Nuphar, and Typha have first occurrence records that
are much younger, ranging from 10 to 7,000 B.P. in New England,
Nova Scotia, and Quebec. The available records and data therefore suggest that these six genera could have survived Wisconsin
glaciation on the Atlantic Coastal Plain and subsequently migrated
northward into New England, Nova Scotia, and Quebec. From the
data of the pollen profiles, it is even more speculative that
these genera might have migrated across the present state of New
York and into northwestern Pennsylvania and northern Ohio. First
occurrence dates in Ohio are considerably younger for some of the
genera, such as a range of 7 to 5,000 B.P. for Myriophyllum, Nuphar, and Sagittaria. A migration route from the Atlantic coast
inland to the Great Lakes region has been discussed both for
indigenous species (Peattie, 1922) and for some recent European,
non-indigenous aquatics (Stuckey, 1966, 1970). A similar postglacial migration route could have been in operation for these
six common genera of aquatic vascular plants.

Records of aquatic vascular plants are scarce in the pollen
profiles from lakes along the southern and southwestern margin of
the Wisconsin glacial ice. Nymphaea, Potamogeton, Sagittaria,
and Typha are known from northeastern Kansas. While this lack of
information may be related to the failure of these plants to be
reported in pollen profiles, the absence of bogs and lakes in
this region of the country is particularly evident. Gates (1976)
has suggested that the ice age climate of the unglaciated continental area of North America during maximum glaciation at about
18,000 B.P. was substantially cooler and drier than of the present
time. This phenomenon may explain the scarcity of potential sites
for the growth of aquatic vascular plants. It appears that with-

out direct evidence the most plausible location for the survival
of aquatic vascular plants was in deep water lakes or backwater
marshes and embayments continually being supplied by meltwater
near the glacial margin in southern Ohio, Indiana, Illinois, and
Iowa. These aquatic vascular plants may have essentially been a
portion of the Mississippi embayment flora which migrated into
the glaciated area via the present Ohio-Wabash River valleys or
northward in the Mississippi River valley and its tributaries.
Invasion into Indiana and Ohio had occurred by 11 to 10,000 B.P.,
and also into Iowa and South Dakota at about the same time.
Younger first occurrences, 9 to 6,000 B.P., occur in northern
Minnesota and eastward into Wisconsin.

Some concern has been given to the idea that these genera of
aquatic vascular plants might have migrated into the glaciated
portion of Ohio and Indiana from populations farther south that
survived on the western side of the Appalachian Mountains in Kentucky and Tennessee. A source for the aquatic vascular plants
in this area may not have existed or awaits more study. Watts
(1970) has stated that the rich aquatic macroflora from a lake in
northwestern Georgia at 22 to 20,000 B.P. shows a marked phyto-
geographic relationship with the modern aquatic flora of soft-
water lakes in New England. His statement suggests an Atlantic
coastal affinity or migration route. In some lakes in Florida
and Georgia the pollen record is absent from about 35 to 8,000
B.P., as reported by Watts (1969, 1972). He has suggested that
the Florida-Georgia region was quite dry because of a lowered
water table during this period. From their southernmost occurrences on the coastal plain in the Carolinas and Virginia at about
17,000-15,000 B.P., it appears more plausible that these aquatic
vascular plants did not migrate around the southern portion or
across the Appalachian Mountains into Ohio and Indiana. Rather,
the Appalachian Mountains have apparently formed an effective
barrier to the colonization of aquatic plants from the southern
portion of the Atlantic Coastal Plain. It was not until the
plants reached the present southwestern New York and Massachusetts area did they migrate westward into the Great Lakes region
or eastward and northward into New England and southeastern
Canada.

Since so few pollen diagrams exist from western or southwestern United States, it is difficult to define the western contribution to recolonization. Our best suggestion is that these a-
quatic plant genera must have survived near the glacial border at
least in the present-day Mississippi valley, perhaps in deep
water lakes or in backwater embayments or marshes of the major
rivers continually being supplied by glacial meltwater as was
stated earlier.

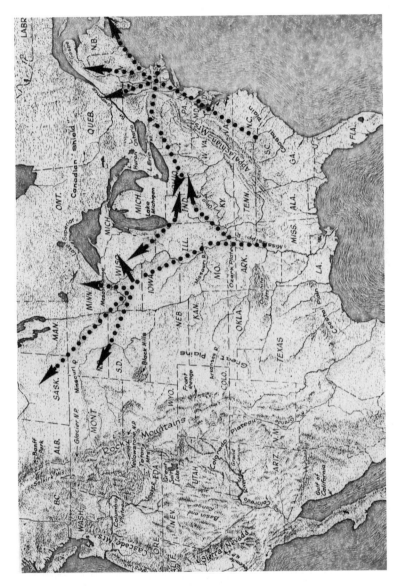

Figure 8.   Suggested migration routes for the return of aquatic vascular plants into the Great Lakes region after late Wisconsin glaciation based on first occurrence records in the pollen profiles for the six genera considered in this paper.

SUMMARY

Two routes for the re-invasion of aquatic vascular plants into the Great Lakes region appear to be evident: the eastern one from the Atlantic Coastal Plain and the southern one from the Mississippi valley (figure 8). These routes correspond to migration patterns that have already been proposed based on the affinities of the present-day flora as derived from the known distributions of selected species. With additional pollen records at both the genus and species levels and with more complete dating of pollen diagrams, a better understanding of the movements of aquatic vascular plants may be obtainable.

LITERATURE CITED IN THE TEXT

Gates, W.L. 1976. Modeling the ice-age climate. Science 191: 1138-1144.

Gleason, H.A. 1923. The vegetational history of the Middle West. Ann. Assoc. Amer. Geographers 12:39-85.

Peattie, D.C. 1922. The Atlantic Coastal Plain element in the flora of the Great Lakes. Rhodora 24:57-70, 80-88.

Stuckey, R.L. 1966. The distribution of Rorippa sylvestris (Cruciferae) in North America. Sida 2:361-376.

Stuckey, R.L. 1970. Distributional history of Epilobium hirsutum (great hairy willow herb) in North America. Rhodora 72:164-181.

Watts, W.A. 1969. A pollen diagram from Mud Lake, Marion County, north-central Florida. Bull. Geol. Soc. Amer. 80:631-642.

Watts, W.A. 1970. The full-glacial vegetation of northwestern Georgia. Ecology 51:17-33.

Watts, W.A. 1972. Postglacial and interglacial vegetation history of southern Georgia and central Florida. Ecology 52:676-690.

LITERATURE USED IN THE PREPARATION OF THE MAPS

1. Brush, G.S. 1967. Pollen analysis of late-glacial and
        postglacial sediments in Iowa, pp. 99-115. In E.J.
        Cushing and H.E. Wright, Jr., eds. Quaternary Paleo-
        ecology. Yale University Press, New Haven and London.
        433 pp.

2. Cushing, E.J. 1967. Late-Wisconsin pollen stratigraphy and
        the glacial sequence in Minnesota, pp. 59-88. In E.J.
        Cushing and H.E. Wright, Jr., eds. Quaternary Paleo-
        ecology. Yale University Press, New Haven and London.
        433 pp.

3. Davis, M.B. 1958. Three pollen diagrams from central Mass-
        achusetts. Amer. Jour. Sci. 256:540-570.

4. Davis, M.B. 1960. A late-glacial diagram from Taunton,
        Massachusetts. Bull. Torrey Bot. Club 87:258-270.

5. Department of Mines and Technical Surveys. 1960. Part 1.
        A palynological study of post-glacial deposits in the
        St. Lawrence lowlands. Contributions to Canadian Paly-
        nology, No. 2. Geol. Surv. Canada Bull. 56. 1-22 pp.

6. Durkee, L.H. 1971. Pollen profile from Woden Bog in north-
        central Iowa. Ecology 52:837-844.

7. Fries, M. 1962. Pollen profile of late Pleistocene and re-
        cent sediments at Weber Lake, northeastern Minnesota.
        Ecology 43:295-308.

8. Gooding, A.M., and J.G. Ogden, III. A radiocarbon dated
        pollen sequence from the Wells Mastodon site near
        Rochester, Indiana. Ohio Jour. Sci. 65:1-11.

9. Gruger, J. 1973. Studies on the late quaternary vegetation
        history of northeastern Kansas. Bull. Geol. Soc.
        Amer. 84:239-250.

10. Hadden, K.A. 1975. A pollen diagram from a postglacial peat
        bog in Hants County, Nova Scotia. Can. Jour. Bot. 53:
        39-47.

11. Harrison, W., R.J. Mallory, G.A. Rusnak, and J. Terasmae.
        1965. Possible late Pleistocene uplift, Chesapeake
        Bay entrance. Jour. Geol. 73:201-229.

S.J. VESPER AND R.L. STUCKEY

12. Jelgersma, S. 1962. A late-glacial pollen diagram from Madelia, south-central Minnesota. Amer. Jour. Sci. 260:522-529.

13. Kapp, R.O., and A.M. Godding. 1964. A radiocarbon-dated pollen profile from sunbeam Prairie Bog, Darke County, Ohio. Amer. Jour. Sci. 262:259-266.

14. Livingstone, D.A., and B.G.R. Livingstone. 1958. Late-glacial and postglacial vegetation from Gillis Lake in Richmond County, Cape Breton Island, Nova Scotia. Amer. Jour. Sci. 256:341-359.

15. Martin, P.S. 1958. Taiga-tundra and the full-glacial period in Chester County, Pennsylvania. Amer. Jour. Sci. 256:470-502.

16. Ogden, J.G., III. 1959. A late-glacial pollen sequence from Martha's Vineyard, Massachusetts. Amer. Jour. Sci. 257:366-381.

17. Ogden, J.G., III. 1963. The Squibnocket Cliff peat: Radiocarbon dates and pollen stratigraphy. Amer. Jour. Sci. 261:344-353.

18. Ogden, J.G., III. 1966. Forest history of Ohio. I. Radiocarbon dates and pollen stratigraphy of Silver Lake, Logan County, Ohio. Ohio Jour. Sci. 66:387-400.

19. Ritche, J.C., and B. DeVries. 1964. Contributions to the Holocene Paleoecology of Westcentral Canada. Can. Jour. Bot. 42:677-692.

20. Scofield, W.B., and H. Robinson. 1960. Late-glacial and postglacial plant macrofossils from Gillis Lake, Richmond County, Nova Scotia. Amer. Jour. Sci. 258:518-523.

21. Sears, P.B., and M. Bopp. 1960. Pollen analysis of the Michillinda peat seam. Ohio Jour. Sci. 60:149-154.

22. Shane, L.C.K. 1975. Palynology and radiocarbon chronology of Battaglia Bog, Portage County, Ohio. Ohio Jour. Sci. 75:96-102.

23. Sirkin, L.A. 1967. Late-Pleistocene pollen stratigraphy of western Long Island and eastern Staten Island, New York, pp. 249-274. In E.J. Cushing and H.E. Wright, Jr., eds. Quaternary Paleoecology. Yale University Press, New Haven and London. 433 pp.

24. Terasmae, J. 1967. Postglacial chronology and forest history in the northern Lake Huron and Lake Superior regions, pp. 45-58. In E.J. Cushing and H.E. Wright, Jr., eds. Quaternary Paleoecology. Yale University Press, New Haven and London. 433 pp.

25. Walker, P.C., and R.T. Hartman. 1960. The forest sequence of the Hartstown Bog area in western Pennsylvania. Ecology 41:461-474.

26. Watts, W.A. 1969. A pollen diagram from Mud Lake, Marion County, north-central Florida. Bull. Geol. Soc. Amer. 80:631-642.

27. Watts, W.A. 1972. Postglacial and interglacial vegetation history of southern Georgia and central Florida. Ecology 52:676-690.

28. Watts, W.A., and R.C. Bright. 1968. Pollen, seed, and mollusk analysis of a sediment core from Pickerel Lake, northeastern South Dakota. Bull. Amer. Geol. Soc. 79: 855-876.

29. West, R.G. 1961. Late- and postglacial vegetational history in Wisconsin, particularly changes associated with the Valders readvance. Amer. Jour. Sci. 259:766-783.

30. Whitehead, D.R., and E.S. Barghoorn. 1962. Pollen analytical investigations of Pleistocene deposits from western North Carolina and South Carolina. Ecol. Monogr. 32: 347-369.

31. Wright, H.E., Jr., A.M. Bent, B.S. Hansen, and L.J. Maher, Jr. 1973. Present and past vegetation of the Chuska Mountains, northwestern New Mexico. Bull. Geol. Soc. Amer. 84:1155-1180.

32. Wright, H.E., Jr., T.C. Winter, and H.L. Patten. 1963. Two pollen diagrams from southeastern Minnesota: Problems in the regional late-glacial and postglacial vegetational history. Bull. Geol. Soc. Amer. 74:1371-1396.

INDEX

Abies, 112
  pollen, 10
Acer pollen, 3, 4
Aclistochara, 181
Alder, see Alnus
Alethopteris, 151, 152, 153
Algae
  charophytes 173-194
  chasmolithic 196
  effects of boring algae
    196-197
  endolithic 195-204
  epilithic 196
  formation of micrite
    envelopes 197, 199-204
  phytoplankton 248-280
Alisporites, 170
Allegheny series, 153
Alnus pollen, 73
Altonian substage, 5
Ambrosia pollen 5
Andreaeales, 96
Andrews Bald 240
Ankistrodesmus 272
Ankryopteris, 137
Annularia, 137, 152
Appalachian Mountains, 10
Aquatic vascular plants 283-296
  recolonization in Great
    Lakes region 284-296
  migration routes into Great
    Lakes region 293-295
Archaeology, 209-227
Archidiales, 103

Archlamydocarpon,
  megasporangium, 82-83
  spore, 82-83
Armeria pollen, 34
Artemesia pollen, 5, 12
Artisia, 151
Ash, see Fraxinus
Asolanus, 137
Asterionella 272
Asterophyllites, 137, 152
Asteroxylon, 96, 98
Ausable focus 217
Bachelor Sandstone, 189
Baiera, 165
Barney Lake, 17
Battaglia Bog, 15
Beaver Island, 16, 17
Beech, see Fagus
Bell Shale 189
Betula pollen, 4, 13
Birch, see Betula
Bones, fossil, 34, 35
Bosmina, 274
Brevispirifer, 188, 191
Bryales, 105
Bryidae, 103
Bryoxiphiades, 100
Bull Brook site, 35
Buxbaumiales, 100
Buzzard's Bay
  ice, 34
  morraine, 34
Calcification of algal
  filaments, 195-204

301